THE NEXT INDUSTRIAL
REVOLUTION

THE NEXT INDUSTRIAL REVOLUTION
Reviving Industry through Innovation

ROBERT U. AYRES

BALLINGER PUBLISHING COMPANY
Cambridge, Massachusetts
A Subsidiary of Harper & Row, Publishers, Inc.

Copyright © 1984 by Ballinger Publishing Company. All rights reserved.
No part of this publication may be reproduced, stored in a retrieval system, or transmitted in any form or by any means, electronic, mechanical,
photocopy, recording or otherwise, without the prior written consent of
the publisher.

International Standard Book Number: 0-88410-885-6

Library of Congress Catalog Card Number: 83-21394

Printed in the United States of America

Library of Congress Cataloging in Publication Data

Ayres, Robert U.
   The next industrial revolution.

   Includes bibliographical references and index.
   1. Technological innovations—United States.
   2. Industry and state—United States.   3. United States—
Economic policy—1981-   I. Title.
HC110.T4A97 1984   338'.06   83-21394
ISBN 0-88410-885-6

CONTENTS

LIST OF FIGURES

ix

LIST OF TABLES

ACKNOWLEDGMENTS

This work was financially supported in part directly by the Department of Engineering and Public Policy at Carnegie-Mellon University (C-MU), and in part by a grant by Westinghouse Electric Company to the Productivity, Human Resource, and Societal Implications of Robotic and Advanced Information Systems Program at C-MU. I should add that the decision to allocate a small amount of Westinghouse grant funds to this project (mainly for secretarial services) was decided upon by the University without consultation with Westinghouse.

Past and present colleagues who have read and critiqued various drafts of this book have made many valuable suggestions and saved me from numerous mistakes. If I have declined to be rescued from sin or error in some instances, the fault is entirely my own. Individuals who have taken particular trouble in responding to my request for advice and assistance on substantive matters include Dr. Robert Aten of the U.S. Department of the Treasury; Dr. James B. Austin (ret), U.S. Steel Corporation; Dr. Vary Coates, Coates, Inc.; Dr. John Fearnsides of MITRE Corporation; Ms. Carol Franco of Ballinger Publishing Company; Dean Robert Kaplan of the Graduate School of Industrial Management (C-MU); Dr. Steven Miller of the Department of Engineering and Public Policy (CM-U); Professor Granger Morgan of the Department of Engineering and Public Policy (C-MU);

xiii

Mr. Thomas Murrin of Westinghouse Electric Corporation; Mr. Michael Naylor of General Motors Corporation; Mr. William Persen of Business International Inc.; Dr. Ronald Ridker of the World Bank; Professor Thomas Saaty of the Graduate School of Business, University of Pittsburgh; Professor William Schultze, Department of Economics, University of Wyoming; Mr. Albert Sobey of General Motors Corporation; Professor Peter Stearns of the Department of History (C–MU); Dr. Wilbur Steger of CONSAD Research Corporation; Dr. Martin Stern, consulting physicist.

My wife Leslie Ayres and my father John Ayres also performed valuable service in editing and correcting several drafts. Finally I wish to thank Ms. Catherine Hill of the Department of Engineering and Public Policy for her cheerful and tireless persistence in preparing the manuscript.

TECHNOLOGY AND ECONOMIC GROWTH

INTRODUCTION TO PART I

In the 1980s the United States has experienced a very deep recession, by far the worst since the mid-1930s. No consensus exists as to the causes or the cures. The immediate problem was acknowledged to be high interest rates, but it was not clear whether rates were high because of enormous expected future federal budget deficits (as Wall Street asserted) or because the financial markets still expected a renewal of inflation (as the Reagan administration insisted), or both. Although the picture has brightened, the deeper problems are not altered by such changes in the financial weather.

The long-term problem is a significant slowdown of U.S. economic growth since 1974 and an even more pronounced decline in international competitiveness vis-à-vis Japan and other countries of the Pacific basin. The most common explanation among economists and businessmen at least, has been that the U.S. slowdown is a transient adjustment problem. This view puts the primary blame on rapidly increasing oil prices (since 1973–74), inflation, and excessive government regulation. Inflation and declining productivity are generally seen by conservative groups as a consequence of faulty government policies, including high taxes, growth of the welfare state, inconsistent monetary policy and Keynesian economic policies, resulting in deficit spending. Some also blame excessive union power.

One conservative policy prescription has been supply-side economics, which entails sharp cuts in personal and corporate income

3

tax rates, deregulation, and cuts in government spending. Optimists asserted that resulting increased economic activity would generate enough additional immediate tax revenue, despite lower rates, to keep down the looming deficits. Unfortunately this prescription for a painless cure has thus far failed, most likely because interest rates remained extremely high, shutting off consumer spending and business recovery. The tax cut may well have been counterproductive because it created an enormous perceived revenue gap that scared lenders and forced the Federal Reserve Board to tighten the screws on credit to compensate.

Other explanations for the "productivity problem" (really a problem of competitiveness) stress a variety of causal factors. Some groups blame the excessive U.S. preoccupation with military adventures such as the Vietnam War, and with nuclear hardware, for diverting scarce resources away from an increasingly threadbare civilian economy. Another popular view is that the United States has lost its technological lead simply because other countries could easily borrow our technological successes while avoiding our mistakes. A related theory puts the blame on a slowdown in U.S. investment in research and development (R&D) since the mid-1960s. The high cost of energy is the villain in another view. "Bad management" is still another theme—one that is growing in popularity. It puts the blame for our economic malaise on preoccupation of managers with short-term profits and mergers at the expense of long-term growth.

Each of these theories has some virtues, but most are ad hoc, in that they attribute specific problems to specific causes (such as the oil price increase) that are themselves left unexplained. They fail to explain all that needs to be understood, including the reasons why the United States has done so poorly relative to other countries with more severe handicaps. In contrast to the many ad hoc explanations, the social sclerosis theory, put forward by Mancur Olson, attributes the problems of our economy to the vast growth of special-interest groups in our complex, long-lived political democracy. This apparently explains the differences between the United States, and Germany or Japan, for instance, where World War II wiped the institutional slate clean and permitted the creation of innovative and effective new institutions. Despite its virtues, however, even the sclerosis theory does not give adequate attention to a critical factor: the special role of technology and technological innovation.

A more satisfying, if still incomplete, explanation of the recent deterioration of U.S. economic performance and international competitiveness is provided by the technoeconomic life-cycle model. For the last century or more, innovative new products have tended to be introduced and produced initially in the United States because it was the biggest, richest market for both consumer and producer goods. Examples include the telephone, camera, sewing machine, automobile, washing machine, vacuum cleaner, radio, record player, air conditioner, tape recorder, television, video cassette recorder, and personal computer. For the same reason, the United States has also been the primary market for most new capital goods, from electric generators to farm tractors, machine tools, and computers.

As demand for the new product grows, the basis for competition among producers gradually shifts from relative performance to relative price. The advantage moves from the innovator to the low-cost manufacturer. This evolution requires larger scale production to exploit economies of scale. This in turn requires standardization of both product and of production technology. When the production technology is sufficiently standardized—as it is today in the so-called smokestack industries—it becomes easily transferable and the optimum location for mass production facilities shifts away from the United States to countries with lower capital labor or material cost. The United States has indeed provided the initial large market for most major new consumer products introduced since the late nineteenth century.

The emigration of production (and jobs) to other countries is visible in the case of textiles and shoes, cameras, watches, consumer electronics, motorcycles, autos, steel, bulk chemicals, and many other standardized, or commoditylike, products. As the jobs are transplanted, so is the purchasing power of the workers. Thus, unless the immigration of older, sunset industries is more than compensated for by fast growth in newer, more dynamic, sunrise industries, the United States' economic stagnation is inevitable and irreversible. It cannot be reversed by tax policies to increase capital formation per se, because the liquid capital—being mobile—will also tend to move to locations abroad where total returns on investment are higher. Capital already invested in plant and equipment, however, is illiquid and immobile. So is much of the labor force, under present conditions. Because of language barriers, costs, and immigration rules, workers cannot move freely from country to country in search of employment. Except for the young and unattached, workers cannot move freely from city to city even within the United States. Mortgages and family ties constitute very effective barriers to mobility.

The cycle of economic stagnation (notwithstanding any short-term recovery) can be reversed only by the conversion of existing capital and labor to new purposes: the creation of new and innovative businesses located in the United States. Thus the productivity problem is ultimately a technological one. The rate of technological innovation in the Unites States—once taken for granted—has slowed down sharply in recent years. In contrast, the rate of technological innovation has sharply accelerated in Japan. On the other hand technological innovation has also lagged in Europe, especially Germany and Switzerland, since the 1950s.

The phenomenon of industrial maturation, with its associated processes of standardization of products and overspecialization of capital equipment and labor, is crudely analogous to the hardening of putty.[1] Much of the productive capital and the labor force in the United States has gradually become immobile and inflexible, partly as a consequence of the natural processes of technoeconomic evolution, partly because of the growth of special-interest groups, as emphasized by Olson's sclerosis theory, and partly because of short-sighted public policies.

Fortunately the information-intensive technologies that are leading the third industrial revolution offer some potential relief from the rigidity of traditional mass-production technology. Whereas the standard approach to minimizing manufacturing cost is to develop highly specialized machines capable of producing one and only one product in enormous quantities, the next generation of factories may gain a competitive advantage from their inherent flexibility. Such a factory may consist of a collection of comparatively general-purpose machine tools, fed by robots, controlled by computers, and capable of producing a wide range of parts within a geometrical family either simultaneously or in sequence. Flexible automation of this kind is no bar to technological innovation. For the United States the question is whether we can shift from the old mass-production orienta-

---

1. Economists use the analogy of "putty" and "clay" to denote mobile vs. immobile (or inflexible) capital and labor.

tion to the new mode of flexible batch production before it is too late for us to compete successfully in world markets.

Public policies and related institutional changes that have contributed to inflexibility are many and diverse. They range from bureaucratically administered government regulations that were originally designed to curb the excesses of competition but that may combine to inhibit the growth itself, to building codes and the proliferation of job classifications and work rules that have resulted from several decades of collective bargaining. The inflexibility that results from financial illiquidity—a consequence of three decades of rapidly increasing and mostly private debt—is another aspect of the problem.

The slowdown in technological innovation is partly due to immobility or inflexibility of capital and labor resulting from a natural process of industrial maturation, and partly due to creeping protectionism in various guises, which has insulated U.S. industry from the need to change. It is also partly due to regulatory and institutional rigidities created by our society. And finally it is due to simple neglect: public support for, and investment in, science and technology has declined significantly since the mid-1960s.

The two crucial questions are, Why has this pattern of adverse changes occurred? And can public policy provide a cure? These are the critical economic and political issues now facing the United States. The first question requires an extended discussion of the mechanisms that control technological change, and the major features of technological change in the past and the near future. The second question is reserved for Part II.

Whether there is a policy cure depends on how the ailment is diagnosed. Effective political action on the necessary scale requires broad public agreement as to the nature of the problem—something that does not exist. The Reagan administration sees the Soviet Union as the major threat to the United States, perceiving the threat in conventional military terms and measuring it by numbers of tanks, ships, missiles, and nuclear warheads. An equally severe yet subtler challenge is posed by Japan and East Asia. It is mounted in economic and technological terms. The battle will be for dominance in the key industries of the future: computers, telecommunications, robotics, and biotechnology. It is not good enough to say, "Yes, technology is important, but the first priority for the federal government is lower taxes and increased military spending—even at the cost of cutbacks in support for science and education." The contrary is true. It would be nice to cut taxes and perhaps to upgrade certain aspects of our military capabilities, but both U.S. economic strength and U.S. military strength now rest almost exclusively on a weakening technological base. If this technological base continues to erode much longer, neither fiscal and monetary tinkering nor deployment of the MX missile will provide security. The first priority of U.S. national policy, for the next two decades, must be to rebuild our technological base.

Obviously national survival must always be the first priority, and national survival *is* very much at stake. But the weakness of the U.S. military establishment is structural and cannot be cured by simply increasing the budget of each service, all else remaining unchanged. Emphasis on procurement—especially of obsolescent weapons systems such as the aircraft carriers and the B1 bomber—seems badly misplaced. The recent package of military counterproposals by Robert McNamara, Cyrus Vance, and Elmo Zumwalt deserves close attention. On the other hand the technological base for future weapons should receive greater priority than even the Reagan Administration has given it.

THE FUTURE OF DEMOCRACY

West Germany and Japan are dynamic, stable, and democratic countries today because of the enlightened postwar policy of the United States after World War II, and because the successful U.S. political and economic system seemed worthy of imitation. Unfortunately, since the mid-1960s neither the international behavior nor the domestic economic performance of the United States has been a model for others to imitate out of respect or admiration. Increasingly, U.S. policies seem to depend on inducing fear in both allies and potential enemies rather than on seeking opportunities for mutual benefit through trade and science.

For the United States, being "number one" is not just a matter of national pride. It is also motivated by a strong and widely shared desire for democracy and liberty to survive and thrive in the face of continuing totalitarian challenges.

Many will argue that we have nothing to fear from democratic, capitalistic Japan surpassing the United States economically, as seems almost sure to happen by 1990, if not sooner. Very strong nonnationalistic arguments exist for not letting this happen.

If Japan could carry the banner for political democracy in the world as well as it sells Toyotas and Sonys, there would be less cause for concern. But Japanese policy since World War II has been totally nationalistic and beggar-thy-neighbor. The reasons—awareness of very limited resources and deep-seated fear of failure—may be strong, but the result to be feared is the potential instability of a multipolar world of shifting alliances, not unlike the situation in Europe in the late nineteenth century. A rich Japan will inevitably rearm itself (as the United States has been urging it to do) and will be less likely than now to see itself as a protectorate or even an ally of the United States. Arms control is difficult to negotiate between two great powers that are almost evenly matched; it will be vastly more difficult— probably impossible—in a multipolar situation.

Worse, a nation that counts its power in obsolete terms and thus believes itself to be strong when it has become weak is extremely vulnerable. This happened to Britain at the end of the nineteenth century. It could easily happen to the United States at the end of the twentieth.

Quite apart from geopolitical factors, a declining country suffers severely in morale and self-image. When citizens cease to feel pride in their common heritage and cease to pull together on behalf of common goals, they are much more likely to find reasons to squabble and quarrel among themselves. Internal class conflict has corroded Great Britain into "little England" and a collection of increasingly separatist provinces. The considerable progress Americans have made in the past 30 years in overcoming the bitter residue of slavery and race discrimination could quickly be undone if the engine of economic growth fails to generate new wealth and to create new opportunities. For all these reasons it is important for us to face the present challenge squarely and respond positively to it.

It would be safer and pleasanter to live in a world without nationalism, under a humane, democratic, and effective world government. The enormous sums now spent on military hardware could be better spent on building roads and railroads, irrigating deserts, reforesting eroded hillsides, damming rivers, and exploiting the oceans and the asteroids. However, since the UN organization does not offer the basis of an effective world government, and since most of the world is governed by totalitarian regimes, a somewhat nationalist approach seems unavoidable, however regrettable it might be.

1 THE DECLINE OF THE UNITED STATES AND THE MYTH OF ECONOMIC INVULNERABILITY

THE U.S. STANDARD OF LIVING

According to the usual measures of societal wealth and well-being, Americans are as well off today as ever before in history. Per capita disposable income (after taxes are subtracted) reached an all-time peak in 1981. This measure has risen almost continuously from a low point in 1933. Real per capita income, after taxes, has grown by close to 340 percent, with the most rapid sustained growth occurring from 1960 through 1979.

The slowdown beginning in 1980 was not unique. A much deeper and more protracted setback occurred after 1944, the peak year of World War II. Per capita disposable income did not ascend to the 1944 level again until 1953 and did not exceed it until 1955. Why should we be unduly concerned today?

The answer can be found in another statistic. The real per capita after tax earnings of American working people (excluding farmers) has been declining since the all-time peak year of 1972, as Figure 1–1 shows. By the end of 1981, when the U.S. Department of Labor stopped compiling the figures, American nonfarm workers had 17 percent less spendable income than they had in 1972 and less than at any time since 1959. (Farmers' incomes, too, were sharply depressed.) The declining trend will probably continue for a number of years to come, because of social security tax increases now on the

11

Figure 1-1. Trends in Real Income, 1947–1981.

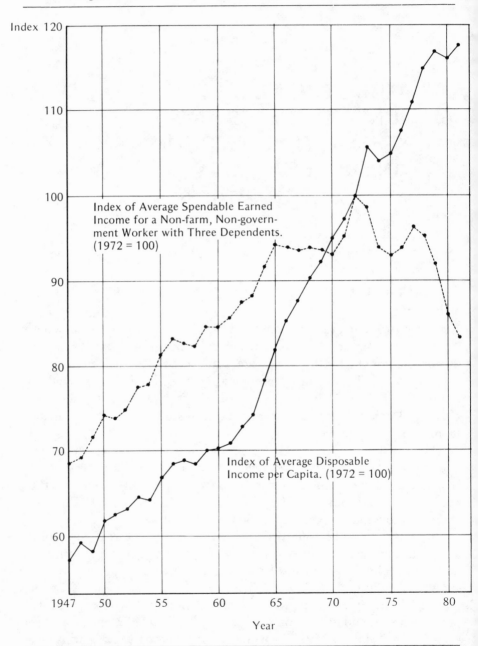

Source: Department of Commerce.

books, unless the U.S. economy exhibits sustained growth at a spectacular and very unlikely rate during the coming decade.

As a measure of standard of living, per capita disposable income refers to the aggregate income, after taxes, of *all* Americans. This category includes both earned and unearned income, of young and old, rich and poor. The other measure, real per capita after tax earnings of American workers, applies only to *earned* income. Obviously, if total income rose while earned income declined, the unearned component accounted for all the increase and more. The unearned portion includes social security payments. And it is only those who are employed—the workers—who pay social security taxes, while it is only nonworkers who receive the benefits of social security. Besides transfer payments, which include federal government pensions, Medicare and Medicaid, welfare, and veterans' benefits, as well as social security, the unearned component also includes rents, interest, and dividends.

All transfer payments come indirectly from the taxpayers. They are paid for, essentially, out of current tax revenues. Because transfers in the aggregate have risen so rapidly, taxes have had to follow suit—outstripping the increase in earned income, from which, in the last analysis, most tax revenues must be extracted. And although most transfer income is also taxable, applicable tax rates tend to be quite low. Thus the working people of this country have suffered a more or less continuous fall in their standard of living for a decade. The causes of the "tax revolt" mandating limitations on real estate taxes and government spending are only too clear. Only the dramatic rise in two-income families has to some extent blurred the impact.

The prospects for young people and disadvantaged minorities entering the labor market are even more adversely affected than for others. Jobs are much scarcer today than they were in the 1960s. Promotions are slower. The American dream of a single-family home ownership has sharply receded. Pensioners, on the average, may be a little better off than they were a decade ago—no doubt they deserve to be—but the rest of us are not. What has gone wrong?

INFLATION AND UNEMPLOYMENT

Capitalist economies since the first industrial revolution have suffered from periodic cycles of rapid growth and prosperity followed by stagnation or recession. There is a very pronounced short cycle, averaging about 3½ years from peak to peak, with several possible longer cycles superimposed on it. The short cycle, sometimes characterized as an inventory-adjustment cycle, has been extensively studied by a number of influential economists during the 1920s and 1930s. It starts from a demand upswing, which leads to demand saturation, overproduction, and inventory buildup, which is followed by production cutbacks, layoffs, and inventory liquidation.

Unemployment generally increases by about 2 percent during the phase of production cutbacks and inventory liquidation; it declines again by a comparable amount during the next phase of inventory buildup. Comparing successive troughs or peaks one notes a long-term trend. In 1945, after World War II, the irreducible level of unemployment was regarded as 2 percent. By the early 1960s this had increased to 4 percent. By the end of the 1970s unemployment remained consistently above 7 percent, even in prosperous periods. It reached 10.8 percent by late 1982 and that figure did not include at least 1.8 million "discouraged" workers, who had stopped looking for jobs and hence disappeared from the official statistics of the U.S. Labor Department. Problems of data collection and interpretation create some ambiguity, but there seems little doubt that the real rate of hard core unemployment is on the rise. This is one of the major symptoms of economic malaise in the United States.

Another variable that fluctuates in rough synchrony with the short business cycle is price inflation. During periods of high demand, bottlenecks occur and prices tend to rise, whereas during periods of slack demand and increasing unemployment, scarcities abate and any impulse toward higher prices is dampened. Commodity prices may actually decrease in a recession. For manufactured goods and wages, a long-term trend is clear. Since World War II, price and wage rises during prosperous periods have *not* been matched by price or wage declines during slack periods. The last period of price declines was 1954–55. The inflationary wage-price spiral, which began around 1966 (probably due to the arms buildup for the Vietnam War), accelerated considerably during the 1970s. The importance of two episodes of rapid petroleum price escalation in 1973–74 and in 1979–80, remains controversial. The rate of inflation reached double-digit levels for the first time in 1974, but dropped sharply during the 1975 recession. Inflation rose again during the Carter presidency, reaching a peak of 13.4 percent in 1980, a fact that contributed greatly to Carter's defeat by Ronald Reagan.

The inflation was not created by Jimmy Carter, however, nor by the overthrow of the Shah of Iran, nor by OPEC, nor even by Lyndon Johnson's oft-maligned "guns-and-butter" policy in 1966–67. Actually, the last federal budget surplus occurred in 1969, due to an income-tax surcharge to pay for the Vietnam buildup. Inflation is most probably due to the fact that in the 1950s and early 1960s American workers became gradually accustomed to a predictable annual improvement in their standard of living. Their rising wage rates were justified by a long period of steadily increasing labor productivity. When the rate of productivity increase slowed down in the late 1960s, workers expectations, voiced as union demands for both wage increases and unlimited cost-of-living adjustments (COLA), continued unabated.

Neither the government nor industry were prepared to resist intense union pressure. The issue was largely decided by the long and expensive United Auto Workers (UAW) strike against General Motors in 1970, which cost the corporation over $1 billion in profits and gave the workers substantial direct wage increases, improved retirement benefits, holidays and other fringe benefits, and unlimited COLA. GM had tried hard, but failed, to win union agreement to several measures to increase productivity. The pattern was quickly adopted by the other manufacturers, suppliers, and many other workers. By the end of 1971 the number of workers entitled to COLA rose from 30 million to 57 million (Serrin 1973), and many non–union-members were among them. Thereafter most of the highly unionized heavy industries, particularly steel and auto manufacturing, chose to avoid further costly strikes by meeting union demands and passing the costs on to consumers by raising prices. Thus inflation was institutionalized. The present troubles of the U.S. automobile industry are at least as attributable to the failure to hold the line on inflationary wage settlements in 1970 and since, as to the vagaries of world energy prices and U.S. energy policy from 1974 to 1980. It is highly significant that in the late 1960s wages in the slowest growing, most heavily unionized industries—steel, automobiles, tires, rails, and petroleum refining—were 19 percent above the average for all manufacturing. By 1980, pushed along by inflation and automatic cost-of-living adjustments, the gap had risen to 42 percent. Steelworkers were paid at a rate over 80 percent above the average manufacturing worker. Not surprisingly, the prices of the products of these industries rose comparably.

The fact that price inflation and unemployment tend to follow the business cycle out of phase with each other results in an inverse correlation between the two. The relationship first observed empirically by economist A.W. Phillips is known as the *Phillips curve*. In the 1950s and early 1960s this relationship was thought to be fairly stable and predictable and became a basis for economic policy. However, since then secular trends in unemployment and inflation have increasingly shifted the curve away from the origin; in fact it has become a spiral (Figure 1-2).

In consequence of the shift, it is harder and harder for governments to disinflate by using monetary tools such as tight money. The reason is that both public policy and union-negotiated job protections have made unemployment progressively less painful. Hence it takes more and more unemployment to cut into aggregate demand. Indeed unemployment may actually have become inflationary (up to a point) as a result of unemployment insurance, welfare, food stamps, and other relief programs that increase government outlays at a time of sagging tax revenues. In fiscal 1983, with unemployment likely to be around 10 percent for much of the year, the federal deficit is projected to be $180 billion, whereas if unemployment dropped to 5.1 percent the deficit would drop to a far more comfortable $28 billion, according to the U.S. Commerce Department. Higher unemployment would result in a still bigger deficit. The revenue gap must be filled by government borrowing—which competes with private borrowing and drives up long term interest rates, hence increasing costs for producers and prices for consumers.

The Federal Reserve Board, which indirectly controls interest rates, expanded credit so fast from 1971 to the end of 1973 under Nixon's presidency that *real* interest rates dropped from an average level of +1 percent to -5 percent. When OPEC started raising petroleum prices in the winter of 1973-74, the Federal Reserve Board reversed its expansionary policy and tightened up on credit, and interest rates went back up. By the end of 1975, real interest rates were again at the +1 percent level. But, at the beginning of 1976, the Federal Reserve Board once again turned on the money spigot, and the real interest rates went back below zero and remained there

Figure 1–2. The Inflation Spiral.

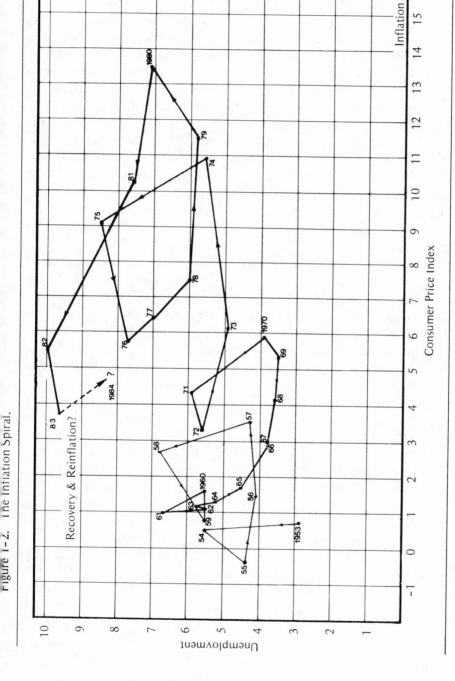

Source: Department of Commerce.

until the end of 1980. All of these gyrations of monetary policy are shown in Figure 1–3.

It is hard to imagine anything more inflationary than negative real rates of interest. The incentive to save and invest obviously diminishes if savings and interest-bearing instruments earn less than inflation takes away. No wonder people have invested in nonproductive collectibles and above all in real estate. An 8 percent or 9 percent mortgage is a bargain if the inflation rate is more than 6 percent or so, because interest payments are tax deductible. In early 1979, 30-year mortgages were still available at around 9 percent—about the same as the then-current rate of inflation. In response to the financial incentives of negative real interest costs, people were bidding up the prices of houses to stratospheric levels. The boom in real estate prices of 1978–79 probably added as much or more to the consumer price index in those years as the second round of petroleum price increases occasioned by the fall of the Shah of Iran and the Iran–Iraq war.

It is not altogether surprising, in view of this dubious record, that since mid-1979, when Paul Volcker became chairman and the Federal Reserve Board finally adopted a monetarist policy for the first time, the Federal Reserve Board moved in the opposite direction and pushed real interest rates to historically high levels. It is not clear at all, however, that the latest course correction by the Federal Reserve Board is the key to long-run economic health. The problems are too deep-seated for monetary policy to solve unaided.

Both unemployment and inflation have been ratcheting up for many years. The relationship between unemployment and inflation oscillates depending on the phase of the business cycle (as shown in Figure 1–2). In the 1950s and early 1960s the magnitudes of the swings were relatively small; the locus of inflation–unemployment (I–U) moved within a small region near the lower left-hand corner of the graph. Since 1968, however, the swings have been getting bigger and both maxima and minima have been growing. If the pattern of the last few cycles is repeated, the noninflationary recovery of 1983 would be followed by a hyperinflationary (election-year) period of prosperity in 1984–85, pushing the next inflationary peak up to 16 percent or so circa 1986. Some pessimists expect even higher rates of inflation during the next peak.

If on the other hand the Federal Reserve Board's extremely tight money policy since late 1979 is a historical break—as the present

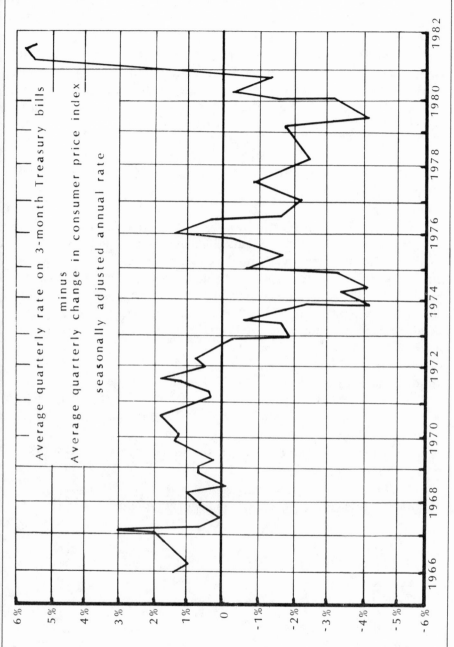

Average quarterly rate on 3-month Treasury bills

minus

Average quarterly change in consumer price index

seasonally adjusted annual rate

Source: Federal Reserve Board, Bureau of Labor Statistics.

high (6 percent) level of real interest rates (Figure 1-1) strongly suggests—then the recovery might abort. In fact, it is still possible that there could be a deep and extended *deflation*, with unemployment rising for several years in a row as overextended businesses and farmers continue to sink under the cost-price squeeze dragging down banks, city and state governments, and even healthy businesses. Either way, Figure 1-2 is telling us that our economy is in trouble and that the trouble seems to be getting worse.

RECENT TRENDS IN PRODUCTIVITY AND TRADE

The United States has led the world in output per employed person for more than a hundred years, having achieved a 50 percent productivity advantage over the United Kingdom as early as 1890. The United States has also consistently enjoyed the highest gross national product (GNP) per employed person, at least among major countries. The average annual rate of productivity growth in the United States, (2 percent per annum) over the century from 1870 to 1965 is roughly similar to those of other industrial nations (Table 1-1). In fact only Japan (2.2 percent per annum) and Sweden (2.1 percent per annum) did better than the United States over that period.

The U.S. GNP per capita actually grew at an unprecedented 3.2 percent per annum rate from 1948 to 1965. But since then there has been a very dramatic slowdown in U.S. growth and a sharp deterioration of U.S. performance relative to other countries. From 1965

Table 1-1. Long-Term Average Annual Rate of Growth of Gross National Product per Capita, 14 Countries, 1870–1965, Average Annual Compound Rate.

Australia	0.9%	Japan	2.2%
Belgium	1.5	Netherlands	1.1
Canada	1.8	Norway	1.8
Denmark	1.9	Sweden	2.1
France	1.6	Switzerland	1.7
Germany	1.8	United Kingdom	1.2
Italy	1.3	United States	2.0

Source: Maddison (1969).

through 1979 U.S. productivity grew at only 1.6 percent per annum. The figure for 1979 was 0.5 percent, and for 1980 it was actually negative (-2.0 percent). Other countries have increasingly outperformed the United States since the 1960s (Table 1-2).

Taken at face value, the success of our major trading partners need not be cause for concern. Their rising standard of living was not necessarily achieved at our expense (and most of them are still not as well-off as we are). International trade yields economic benefits to *all* participants, at least in principle. Why, then, speak of a U.S. decline? The decline is one of comparative performance and advantage, but it is very marked. The nature of the problem is clearer when U.S. trade performance is compared with that of other countries. In the years 1951-1955, U.S. foreign trade was small compared to internal trade, but the balance of merchandise trade was a comfortably positive $4.6 billion per year, even allowing for a $1.2 billion deficit in agricultural products and a $2 billion deficit in energy and raw materials. A subset of technology-intensive products yielded a $5.7 billion surplus, while there was small deficit ($0.9 billion) in products not classed as technology intensive.

During 1964-65 the trends began to change. Both exports and imports started to increase rapidly, but imports increased still faster. In the mid-1970s the volume of international trade suddenly jumped as a flood of "petrodollars" began circulating, but the long-term trends continued. By 1980 the pattern had drastically changed for the worse. The energy and raw materials deficit had soared to $68 billion, partly balanced by a $24 billion surplus in agricultural products and $42.5 billion surplus in the technology-intensive category. But there was also a $21.5 billion deficit in trade in non-technology-

Table 1-2. Ratio of Selected Countries' Gross Domestic Product per Capita to That of the United States, 1960-1980.

	1960	1970	1979 or 1980
United States	100.0	100.0	100.0
Germany	73.3	82.3	87.4 (1980)
France	61.6	75.9	80.0
Japan	31.5	61.9	70.2
United Kingdom	66.5	64.9	58.6 (1980)

Source: Kravis et al., (1981).

intensive products, resulting in an overall $20 billion deficit in merchandise trade. The declining U.S. share of world trade in technology intensive products is shown by Table 1-3.

Moreover most of the 1980 surplus in technology-intensive products came from trade with non–OPEC countries, which had to be financed by long-term U.S. loans. Some of these loans to countries like Mexico, Brazil, and Argentina now appear to be seriously in jeopardy. Trade balances with both Germany and Japan on the other hand have been progressively more negative. West Germany, with a 1980 GNP only 21 percent as big as the U.S. GNP, had a trade surplus in the technology-intensive category that was 50 percent larger in absolute terms than that of the United States. Japan, with a 1980 GNP 34 percent as big as the U.S. GNP, had a trade surplus in this category 67 percent larger in absolute terms. Adjusting for the size of their economies, Germany performed about 7 times as well, and Japan performed 5 times as well as the United States, as exporters of technology-intensive products! Even France outperformed the United States by this measure.

The situation is particularly disturbing when one considers that the United States is running a huge trade deficit with Japan in a wide range of products that were once showpieces of U.S. mass production. The list runs from ball bearings and fasteners like nuts and bolts to motorcycles, sewing machines, typewriters, televisions, and automobiles. As early as 1963, 60 percent of the typewriters and 66 percent of the sewing machines purchased in the United States were made abroad. Some products, like 35mm cameras and most consumer electronics products are no longer manufactured in the United States. And the U.S. automobile industry has given up 26 per-

Table 1-3. World Trade Shares in Technology-Intensive Products.

	1954	1970	1980
United States	35.5	23.1	19.9
United Kingdom	19.0	10.1	9.0
West Germany	17.6	20.4	19.3
France	6.4	7.6	9.0
Italy	2.4	5.6	5.5
Japan	1.8	9.7	14.5

Source: U.S. Department of Commerce, unpublished (1982).

cent of its domestic market to imports, of which 81 percent are made in Japan. When the current Japanese voluntary quota limitation expires in 1984 (unless it is extended), many observers expect the Japanese market share to rise quickly to 40 percent. There is no reason to believe that 40 percent would not eventually become 50 percent, 60 percent, or more.

The basic reason for this Japanese penetration of the U.S. market is that by 1980 the Japanese could produce a $6000 subcompact car and deliver it to the West Coast for $1700 less than Detroit can. According to independent studies by former Chrysler executive James E. Harbour and a group at Harvard Business School, the gap is attributable to a variety of factors: lower parts inventory ($550), better quality controls ($330), lower labor costs and better use of labor ($125), lower absenteeism ($80) and miscellaneous factors ($215) (*New York Times Magazine*, November 14, 1982; Abernathy et al. 1981). Reputedly, General Motors loses around $900 per Chevette sold. Other Detroit-made subcompacts are also unprofitable, even the best-selling Ford Escort. Clearly, barring a technological tour de force not yet visible to outsiders, Detroit will eventually be forced to give up the low-priced end of the market unless the Japanese voluntarily restrain exports or Congress changes the rules. Detroit can only make profits on larger cars that the Japanese do not yet manufacture in large numbers.

U.S. export performance has been a bit better in the category of high-technology industries (including pharmaceuticals, office equipment and computers, electrical and electronic equipment, aircraft and missiles, and instruments and controls). Here, at least, most of the U.S. trade surplus is from trade with other developed countries in the Organization for Economic Cooperation and Development (OECD) or with OPEC. The 1980 surplus in this category was $14 billion. But again, Japan already had an even larger surplus, in absolute terms, and a much higher growth rate. Ominously, the United States is able to export very little to Japan except raw materials and agricultural products.

The problem is exposed in another way as a shift in world market shares of technology-intensive products. In 1954 the United States produced 35.5 percent of the world's exports of such products, as compared to 19 percent for the United Kingdom, 17.6 percent for Germany, and only 1.8 percent for Japan. By 1970 the U.S. share was down to 23.1 percent, and the United Kingdom's was down to

10.1 percent, while Germany's was up to 20.4 percent and Japan's share had risen to 9.7 percent. By 1980 the Japanese share was up to 14.5 percent, at the expense of both the United States and Europe (except France). But the United States was by far the greatest loser of market share in the 1970s.

A case in point is the machine-tool industry, a lynchpin of any modern industrial economy. The United States has been a major machine-tool innovator, builder, and exporter since the late nineteenth century. In the early 1970s U.S. machine-tool builders still exported twice as much as they imported. Exports exceeded imports until 1977, when imports of computer-numerically controlled (CNC) machine tools from Japan soared. By 1982 the balance of trade in machine tools had reversed: imports exceeded exports by more than 2 to 1, and over 50 percent of the machine tools purchased in the United States were manufactured elsewhere, mainly in Japan. The U.S. machine tool industry has lost half its traditional market in the past five years.

THE INNOVATION LAG

The slowdown in innovation in the 1960s seems self-evident, but documentation has not been completely satisfactory. In an attempt to fill the gap I have compiled from many sources a list of nearly 800 major technological innovations in nine categories, by decade, since 1730, Both weighted and unweighted tabulations were prepared. In the weighted version, I gave extra weight to the most important inventions, using a simple scale. Most innovations received a weight of unity but approximately 100 were given a weight of 2 while 15 very important breakthroughs received a weight of 4. Significantly, the weighted and unweighted results were very similar, as shown in Figure 1-4 through Figure 1-7. Evidently, the more important innovations are distributed in roughly the same way as the less important ones. Thus, though I may have given too much or too little emphasis to innovations in one field as compared to others, the resulting distortions do not seriously affect either the inter-country or intertemporal comparisons. The distribution of major innovations by country since the 1930s is shown in the pie charts, Figures 1-6 and 1-7. (Complete supporting data and analysis will be published separately in a journal article).

Figure 1-4. Number of Innovations for Europe, America, and Others, 1730s–1980s.

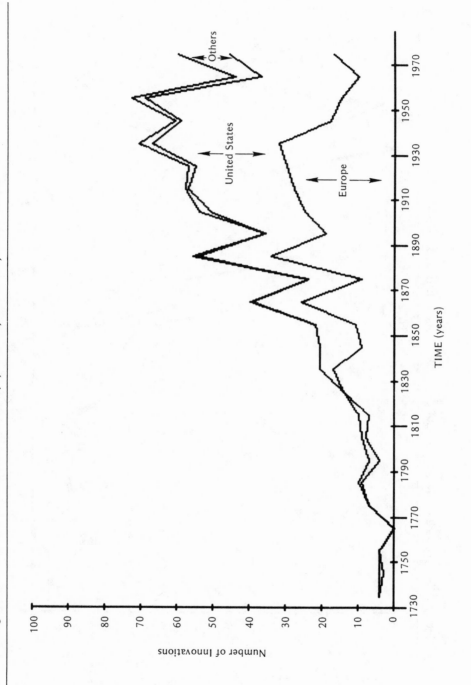

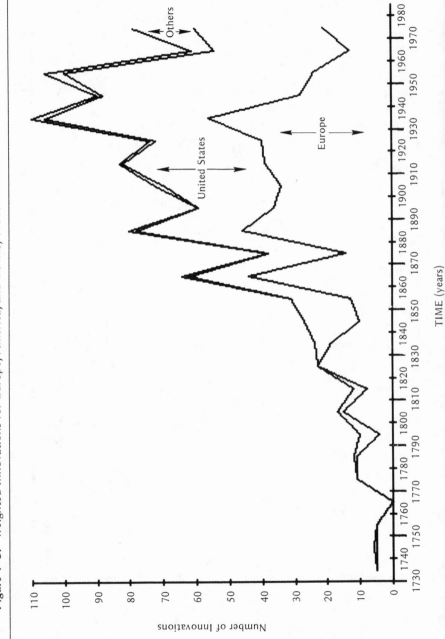

Figure 1-5.   Weighted Innovations for Europe, America, and Others, 1730s–1980s.

Figure 1-6.   Distribution of Innovations, 1940s-1970s.

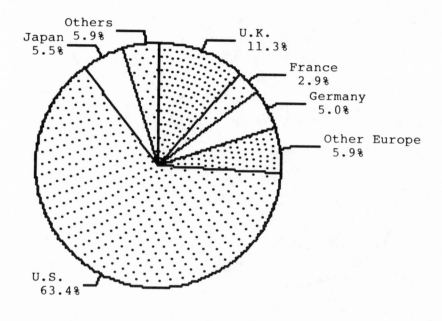

Many interesting results of the study cannot be fully substantiated without more detailed analysis, but several major conclusions seem inescapable. First, the United States has accounted for about half of all major innovations since 1730. The United States surpassed Europe in innovation in the late nineteenth century, having generated more major innovations in every decade since 1930 than all of the European nations combined. Second, U.S. technological dominance peaked in the 1950s and 1960s, though this was mainly due to a sharp decline in European innovation rates, especially for Germany. In fact, the trends for Germany bode ill for the future of the German economy. Third, the innovation rate for the world as a whole in the 1960s was lower than any decade since the 1920s and sharply lower than the rate in the 1930s and 1950s. Fourth, the 1970s' rate of innovation rose slightly for the world, but the U.S. share of the total dropped from 67 percent to 48 percent. On the other hand, Europe (except for Germany) did better in the 1970s than in the 1960s, while Japan has sharply accelerated its rate of innovation since the 1960s.

Figure 1-7. Weighted Distribution, 1940s–1970s.

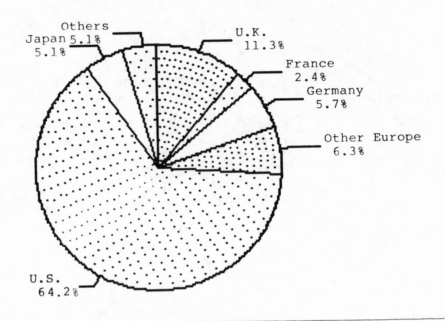

Apart from the evidence provided by the tabulation, anyone who has followed the technological literature over the past two decades must be aware of the large number of important innovations that observers in 1960 generally expected to occur before 1980 but that were aborted or seriously delayed. Here is a partial list:

Synthetic protein from petroleum or lignin wastes
Fuel cells widely used
Magnetohydrodynamic (MHD) power generation
Superconducting electric power transmission
Ground effect machines (e.g., "Hovercraft") widely used
Magnetic levitation for high-speed trains
Vertical Take-Off and Landing (VTOL) aircraft in use
Supersonic and hypersonic civilian airliners
High energy batteries (e.g. zinc-air, zinc-chloride sodium-sulfur, or lithium)
Electric cars widely used in central cities
Aluminum/plastic auto bodies
Stirling cycle engine for trucks and buses

Aluminum chloride (ALCOA) process for aluminum smelting
Large-scale desalinization of seawater
"Picturephone" widely used
Holography widely used
Computer language translation available
Computer-controlled factories
Effective chemotherapy for cancer
Radiation sterilization for food preservation in use
Effective drug therapy for mental illness

Specific reasons for the slowdown in technical progress in these and other fields are easily cited. But the fact remains that there has been an overall slowdown of significant magnitude.

AGING PHYSICAL INFRASTRUCTURE

Infrastructure denotes streets, highways, railroads, canals, harbors, dams, bridges, water and sewer systems, electricity generation and distribution systems, gas distribution, water, and sewer systems, and communications systems. Although electricity, gas, and telephone systems remain relatively sound in the United States, the condition of the rest ranges from substandard to terrible.

The condition of the railroads is symptomatic. In the mid and late nineteenth century there was a tremendous amount of railroad-building in the United States. By the 1920s the system was, by any standards, superb. Virtually every American town was served by one or more rail lines, and service was frequent, efficient, and inexpensive. Because the rail system was so important to the economy, railroads gradually came under heavy regulation by the federal government. As railroads were perceived to be rich monopolies and in some cases behaved with extreme disregard for the public, their rights-of-way were increasingly heavily taxed by towns they served. Railroad unions also devised extremely restrictive and costly work rules to protect jobs by inhibiting automation. Diesel-electric engines had to carry "firemen," freight trains had to haul manned cabooses, and crews had to be changed at fixed mileage intervals regardless of speed or scheduling considerations.

Worse for the railroads, the federal government helped to create and continues to subsidize a powerful competitor: truck transportation using the interstate highway system. Truckers argue that the highway system is self-financing through the federal and state motor fuel tax. This is true in a very narrow sense, but publicly owned rights of way do not pay local real estate taxes, as railroads have had to do. Moreover, truckers have been subsidized by private cars, to the extent that trucks cause disproportionate damage to the roads. Competition from trucks has been economically devastating for the railroads. Facing virtually fixed labor and capital costs and steadily declining demand, U.S. railroads have been forced to raise freight rates (to the extent the regulatory agencies permitted) while cutting back on investment in rolling stock and long-term maintenance of tracks and structures. Because tracks are in poor condition, trains have had to run slower and slower. The Amtrak passenger train between New York and Boston averages only 42 miles per hour and has a top speed around 80—an embarrassing contrast with Japan's 120-mile-per-hour bullet train from Tokyo to Osaka and France's even faster (160 miles per hour) new *tres grand vitesse* (TGV) train between Paris and Lyons. The low level of investment in rolling stock by U.S. railroads has resulted in the virtual disappearance of the U.S. rail car-building industry.

Meanwhile the interstate highway system is also beginning to show signs of wear. The Federal Highway Trust Fund, once an overflowing source of construction funds replenished by the federal gasoline tax, has been tapped to help support urban mass transit without much benefit to the latter. (The condition of urban transit systems needs no comment.) The interstate highway system, though still incomplete, has begun to deteriorate faster than it can be repaired out of the existing revenues. The major cause of the deterioration of the interstate system appears to be intensive use by a fleet of 800,000 tractor-trailers, which now carry the bulk of the nation's freight, but which have hitherto paid a comparatively small share of the highway taxes and user fees.[1]

State and local roads—by far the bulk of the road mileage—are in even worse shape. Many roads were built in the 1920s and 1930s when traffic and vehicles were much lighter than today. They are now wearing out, at an accelerating rate. Patching is no longer sufficient. Hundreds of thousands of road-miles now need to be rebuilt to higher standards than the original construction. Bridges constitute a special problem. There are about 560,000 bridges in the United States, mostly of steel construction and mainly built between 1880 and 1940. Of these, no less than 50 percent are either in need of repair (28 percent) or replacement (22 percent). Apart from chronological aging, the rate of deterioration has sharply accelerated in recent years due to the increased use of salt for purposes of melting snow and ice. Yet during the entire last decade, 1972–1982, only 9,000 bridges were replaced, or 1.6 percent of the total number. In the state of Pennsylvania alone, there are 54,500 bridges, of which at least 4,000 are now closed or restricted and over 7,000 are obsolete and in need of replacement. Clearly the present rate of repair and replacement is far too low (Fenves 1983).

The overall federal investment in public infrastructure (interstate highways, dams, canals and locks, water-supply, sewage treatment) amounts to $1.3 trillion (1981 dollars). Adding state and local investment (streets, highways, water, and sewer) would bring the total up to at least $2 trillion. Documented repair and replacement needs for public infrastructure alone (excluding needed railroad investment) amount to over $900 billion, including $320 billion for highways, $120 billion for dams, $120 billion for sewage treatment, and $50 billion for bridges (Fenves 1983). Yet current annual expenditures for all these purposes by federal, state, and local government amount to barely $54 billion—certainly not enough to compensate for the annual deterioration.

In the past there was some overbuilding—certainly this was true for railroads—and a healthy economy can do without some of the obsolescent or obsolete facilities included in the inventory. Nevertheless it is clear to any observer that most of these asserted needs are real. Moreover we pay for deteriorated roads and other systems in terms of unnecessary wear and tear on vehicles, transportation bottlenecks, detours, and delays (time is money), and accidents, not to mention adverse impacts on quality of life and on the environment.

LIVING OFF CAPITAL

The greatest economic sin to a Victorian Englishman was "to live off capital," meaning to dip into principal, as opposed to spending interest income. As the nineteenth century wore on, more and more well-to-do Britishers invested their capital in gilt-edged stocks or "consols" or Argentine Railway bonds at 3 percent or 3½ percent and lived comfortably on the proceeds. Productive effort was not admired in Britain. The conservative capital investments proved, in the long run, to be the riskiest. The capital the well-to-do English invested so carefully but failed to renew is long since gone—eaten up by taxes, inflation, and various domestic and international disasters.

There are other ways for a nation to live off its capital stock. The simplest is to fail to replace worn-out or technologically obsolescent physical plant and equipment. That the United States has recently been failing to renew its physical capital is painfully obvious. In the private sector the situation is not much better. According to the last *American Machinist* national machine tool inventory, carried out in the years 1977–1979, 34 percent of the U.S. machine tools were more than 20 years old, and only 31 percent were less than 10 years old. Even the United Kingdom had more modern equipment: only 24 percent of its tools were over 20 years old, while for Japan, the comparable figure was 18 percent in the oldest group, with 61 percent of Japanese tools less than a decade old.

The aging of physical capital is due to lack of investment. The immediate excuse for this is that factory capacity utilization in the United States has been declining since the mid-1960s. Utilization reached a cyclic peak level of 91 percent in 1966, with subsequent peaks of 87 percent in 1973 and 85.5 percent in 1979. The 1971 low point was 78 percent, the 1975 low was 73 percent, but capacity utilization dropped below 68 percent in late 1982. For the steel industry, capacity utilization dropped below 50 percent. With so much excess capacity and so little demand, business management is reluctant to invest in new equipment, even if the investment would sharply cut costs at higher production levels. At one time the guiding principle for investment was to be sure to have enough capacity to maximize profits during the next economic upturn. During the 1981–82 recession, however, many firms in economic difficulties deliberately cut back on reserve capacity to cut overhead and reduce their break-even points. Should a sustained recovery occur, much of the added demand will perforce be met by imports.

Inflation itself is another way of depleting capital, since the major institutionalized method of replacing capital in our economy is the investment depreciation allowance. But depreciation allowances are based on the original value of the investment, not the inflated replacement value. Thus depreciation becomes less and less adequate, as inflation heats up. Even though corporate revenues tend to rise with inflation, more of those inflated revenues appear as taxable profits and less is available for capital replacement. Thus inflation results in phony increases in earnings that in turn generate real increases in tax revenues for the government—which are recycled as still more inflation-inducing expenditures and more current consumption.

The increased debt load also creates problems of another sort. Not only do the carrying costs increase as inflation pushes up interest rates, but the burden of short-term debt reduces the liquidity—hence the financial flexibility—of the debtor. The illiquidity of U.S. banks and industry in 1982 was at or near record levels, with only modest improvements through summer 1983. The problem is that one or two large failures could set off a chain reaction. While the Federal Reserve Bank would probably step in, it could only rescue the banks by monetizing the bad debts creating more inflation.

The U.S. social security system, including Medicare and Medicaid, constitutes another example of borrowing, albeit in somewhat disguised form. We as a society have made large commitments for payments to the elderly and the sick that a future generation will have to meet out of its current income. In a true insurance system, money is saved by individuals and aggregated and invested by the insurance companies. The income from these investments limits and determines the repayments to individuals. But social security operates on a pay-as-you-go basis, because past and current social security payments have been set at levels that far exceed the earning power of actual cumulative contributions. Since the system began operating in 1930s, there has been less incentive for people to save for their own retirement years or to provide for medical and other emergencies. In effect we are transferring income from people now working to people now retired. We are not endowing the workers of the present generations with the capital from accumulated savings to keep creating that income when they in turn wish to retire. This is equivalent to living off capital.

The progressive exhaustion of high-quality natural resources such as timber, petroleum, iron ore, copper ore and fossil groundwater, is another case in point. We are also losing significant quantities of valuable topsoil to wind and water erosion. According to industry insiders, the average cost of finding a new barrel of petroleum in the United States in 1981 was $28, very close to the current market value. At this price fewer reserves were discovered than were consumed. This is ominous, for it means that when presently available supplies are used up, costs are likely to take another sharp jump upward. The longer the present period of glut lasts, together with declining prices, the more traumatic the next round of oil price increases will be. As regards groundwater, when the irrigating pumps throughout the Great Plains run dry in a decade or two, there may be no economically feasible alternative source of water for large areas. Less valuable crops will have to be grown. Lost topsoil, too, is probably irreplaceable.

THE "PROBLEM OF PRODUCTION"

When Eisenhower was president in the 1950s, Americans knew for certain that they lived in the richest, most advanced, most powerful country the world had ever seen. The Soviet Union could pose a military threat in Europe, to be sure, but Americans had no equals, or even serious challenges, in the marketplace. It was American industrial prowess after all, that won World War II for the democracies. The war production feats of Kaiser, Chrysler, Ford, Douglas, and Lockheed—to name a few—were legendary. The world wanted to buy products *Made in America* so much that foreign travel, foreign aid, and foreign investment, on a large scale, were essential means of balancing the flow of money. Americans in the 1950s were not interested in buying European or Japanese products, but American corporations were interested in buying up European firms at bargain prices. In those distant days the prospect of serious competition from Europe and Japan seemed almost inconceivable to American business.

Even a decade later, most large-scale American enterprises still seemed quite invulnerable. The perennial number three American auto company, Chrysler, was bigger and more profitable than any foreign-based competitor. General Motors, the Goliath of the industry, had annual profits greater than the total sales volume of all but two or three foreign producers. A similar situation prevailed in most other sectors of the economy. The fact that foreign, especially Japanese, firms were rapidly increasing their share of the world market worried almost nobody at the time.

In a series of influential books published over a period of two decades, John Kenneth Galbraith argued that American capitalism had solved the problem of production and that it was therefore time for the national agenda to shift its focus to other social concerns (e.g., Galbraith 1958). Galbraith's argument rested on the proposition that very large modern corporations control their own environment; they are no longer subject, as small businesses are, to the so-called laws of supply and demand. He restated his basic position many times. For instance,

> The initiative in deciding what is to be produced comes not from the sovereign consumer who, through the market, issues the instructions that bend the productive mechanism to his ultimate will. Rather, it comes from the great producing organization which reaches forward to control the markets that it is presumed to serve and, beyond, to bend the customer to its needs. (Galbraith 1978, p. 6)

In all fairness, there is an element of truth to what Galbraith said. Corporate executives themselves believe in bending the consumer, to a degree. Otherwise why spend hundred of millions of dollars annually on advertising? *But corporations, no matter how large, cannot make people buy products they really do not want.* This has been starkly demonstrated several times in recent years. Galbraith believed the private sector, by virtue of its power over markets, was omnipotent, while the public sector starved for resources to solve social problems. Galbraith's argument was widely heard and believed. The alternative political agenda he advocated was adopted and virtually enacted in President Johnson's Great Society programs in the mid-1960s. Meanwhile corporations were reviled and increasingly regulated for their numerous sins of commission and omission.

One example is worth recounting. Right after the 1973–74 oil shortage, the American public demanded small cars, and all the manufacturers—domestic and foreign—were unprepared. Small-car sales jumped in January 1974 from 36.8 percent of the U.S. market to 48.4 percent. There was panic buying of small cars, while big cars piled up in inventory. All four U.S. producers immediately began converting large-car plants to increase small-car output, and industry experts forecast that small cars would rise to 50 or 60 percent of total sales by 1980. Yet a mere three months later gasoline prices had leveled off and the public had lost interest in small cars. By July 1974 Ford Motors had a 90-day supply of Pintos and had to lay off 2,000 workers at a Pinto assembly plant. General Motors had a 110-day supply of Chevrolet Vegas, while Chrysler had a 105-day supply of Plymouth Valiants and a 113-day supply of Dodge Colts. By contrast, GM had only a 26-day supply of Cadillacs and had to go to double shifts to produce more. The automobile industry offered rebates on the small cars, but inventories continued to rise.

In December 1974 Henry Ford II asked the government to raise the federal gasoline tax by an additional 10 percent to increase the attractiveness of fuel-efficient cars. President Ford's energy advisor, John Sawhill, advocated a much stiffer tax of $1 per gallon to raise gasoline prices in the United States closer to the levels of Europe and Japan. Sawhill soon lost his job, however, and the American automakers took a beating trying to sell more small cars than the public really wanted to buy at that time—or (if current evidence is to be believed) anytime since.

Detroit was naturally willing and eager to sell larger cars, which Congress in its wisdom made attractive by controlling the price of domestic oil. As an afterthought to compensate for this, Congress simultaneously imposed fuel-economy standards on the industry. The manufacturers then found themselves with an unfillable demand for large cars—especially with V-8 engines—and an excess supply of unsellable small cars. In January 1979 Datsun was offering rebates, and Toyota and Honda were selling poorly. Yet there were actually waiting lists for some of Ford's and Chryslers' large car models (see Tucker 1980 for details).

A few months later the balance of forces in the world had shifted again. The Shah of Iran was overthrown by the Islamic fundamentalist revolutionaries of Ayatollah Khomeini, Iran cut back its oil production, gasoline prices doubled in a year, and large cars were again unsalable. Yet at no time from the beginning to the end of this period was the automobile industry able to predict the market at all accurately, still less to reach forward to bend the customer to its needs. Lacking this miraculous ability, the U.S. automotive industry is now in very deep trouble, despite its short-term (1983) profit recovery. In fact, both the companies and the union (UAW) depend upon an indefinite continuation of the present quota limitation on Japanese imports. Without this protection, the Japanese would eventually build up their capacity to compete in every market niche.

In fact Chrysler Corporation has sold a number of its assets and avoided bankruptcy only with the help of $1.2 billion in government loan guarantees. In 1982 Ford Motor Company broke the all-time record for the biggest loss in a single year—over $2 billion. A–M International (formerly Addressograph–Multigraph) and Braniff Airlines became insolvent and sank. As of autumn 1982 several other giant manufacturing firms, including International Harvester, Allis-Chalmers, Massey–Ferguson, and AEG–Telefunken were losing massive sums and teetering on the edge of bankruptcy, having avoided it only because of heroic and unwilling rescheduling of debt by their banks. Smaller firms were going bankrupt quite regularly—at the rate of 20,000 per week—double the rate in 1972.[2] The "problem of production" remains unsolved. Perhaps it would be more accurate to say that the nature of the problem keeps changing, so that old solutions must constantly be rethought.

UNITED STATES PAST VERSUS JAPAN PRESENT

Changes in the United States between 1910 and 1982 are summarized in Table 1–4. Many of these changes contribute directly to current U.S. economic problems. In the long run increasing resource scarcity and international competition may stimulate innovation. On the other hand the high levels of regulation, concentration, and unionization of basic industries are not really conducive to innovation, according to the Olson sclerosis theory. The passing of the baby bulge will absorb some of the current excess labor supply, but only if the U.S. economy continues to grow. But an aging and increasingly dependent population will make saving and capital accumulation more difficult. (Japan will have this problem also, since its population is on average even older than ours.)

Table 1–5 compares the U.S. situation in 1982 with the situation in Japan. It is perhaps clearer from the comparison why U.S. performance has lagged than why Japan has had such good economic performance for the last 30 years. Japan's having no natural resources to speak of evidently has provided a very helpful incentive. But the economic success of Japan is not entirely due to its lack of resources, nor to its high rate of saving and investment. Other key factors include the deliberate Japanese national policy of seeking out, acquiring, and adapting technology developed elsewhere (Lynn 1982). Another critical factor arises from difference in corporate structure and management styles (Ouchi 1980; Hofheinz and Calder 1982). Some major Japanese firms are still run as personal feifdoms by self-made entrepreneurs who started from scratch after World War II. Sony, Matsushita, Honda, and Fujitsu–Fanuc fall into this group.

Table 1-4.   Changes in the U.S. Comparative Advantage, 1910–1982.

Labor:	Skilled labor was scarce in 1910, but most skills are in surplus supply in 1982. The U.S. labor force was then the best educated. Education measured in terms of literacy and ability to calculate is distinct from specialized skills involving particular tools, such as carpentry, steam fitting, welding, and so on. It was also the most mobile in the world. It is no longer best educated or most literate, though it is still relatively mobile. Unionized labor, however, is now very expensive, inefficient, and immobile. From 1978 through 1981, U.S. workers were on strike, 3,280 days per 1,000 employees.
Resources:	The United states was self-sufficient in virtually all minerals and was a major exporter of petroleum in 1910. It is now an importer of petroleum and many critical minerals. (On the other hand, the United States is still the world's greatest food producer and exporter.)
Capital:	In 1910 U.S. infrastructure, plant, and equipment were new. They are now relatively old and energy-inefficient.
Technology:	The American System of Production was still unique to the United States in 1910; it has long since been adopted and improved upon worldwide. Technology of production for most standard products is now available worldwide, on a turn-key basis. U.S. industry, unfortunately, is still energy-intensive, based on decades of cheap energy. More energy-efficient technology is only now beginning to be utilized.
Science and Education:	In the late nineteenth century, the United States was not as advanced as Europe in basic scientific research. Its great strength was widespread literacy and popular interest in technology, especially practical applications. By the midtwentieth century the United States had attained world scientific leadership, but the secondary education system has weakened.
Market Orientation:	The U.S. market was then and is now the world's largest, but other markets are growing faster. U.S. manufacturers are not yet oriented to selling in world markets; for instance Americans do not learn foreign languages and most are unfamiliar with the metric system of weights and measures.
Incentives and Institutions:	In 1910 there were few regulations (except for antitrust laws). U.S. industry is now much more heavily regulated and a great many disputes are routinely referred to the courts—adding enormously to social overhead costs. Tax incentives in the United States favor consumption more than investment. Managers in the United States are expected to maximize results in the very short run, thus neglecting long-run strategies.

Table 1-5. Comparison: United States versus Japan, 1982.

Labor:	Japanese labor is well educated, disciplined, and highly motivated. Skilled labor has been and is scarce in Japan, due to rapid economic growth and low birthrate. From 1978 through 1981 Japanese workers were on strike only 170 days per 1,000 employees.
Resources:	Japan imports virtually all natural resources, including much of its food.
Capital:	Japan's plant and equipment are new and energy efficient. Japan's savings rate is very high. (Some economicts emphasize that the savings rate in the U.S. has dropped as the dependency ratio has risen so that retired people are being supported by fewer working people than in 1910. However, the dependency ratio in Japan is even higher than that in the United States.)
Technology:	Japanese technology is now as good as U.S. technology in most fields. The Japanese are better at assimilating new ideas from elsewhere.
Science and Education:	Japanese science is adequate and improving rapidly but not yet world leading. The basic (especially secondary) education systems is better than that of the United States.
Market Orientation:	Japanese industry is attuned to world markets. Japanese firms frequently develop products primarily for foreign markets.
Incentives:	Japanese industry is less heavily regulated, and government is much more supportive. Tax incentives favor investment and saving. There are very few lawsuits. Managers plan for long-term results. The reasons for this are partly institutional and partly cultural.

But among the more established older Japanese firms, the system of management is based on collective decisionmaking, shared responsibility, and informal control. Employees are hired straight from schools or universities, and job-hopping is very rare. (The average rate of employee turnover in Japan is 4 percent per year, compared to 12 percent in Europe and 26 percent in the United States (Ouchi 1980).)

Company and collegial loyalty in Japan is high. Careers are unspecialized, promotion is slow, and salaries are based essentially on seniority, not on job title. Whereas Japanese firms are slow to reach strategic decisions because of the collective nature of the process, they are notoriously quick to implement any decision once made. This seems to be because there is less rivalry for credit—or to escape blame—and less need for further explanation and negotiation among the parties involved. There are also many fewer levels of management in Japanese firms than their U.S. counterparts. For instance, it was discovered when Toyota and GM embarked on a joint venture to build small cars in California that Toyota has five fewer hierarchical levels than GM.

American firms, in contrast, are based on a far more individualistic system of management. Responsibility tends to be fragmented, not shared. Managerial employees move easily from company to company, being more concerned with their own careers than the good of their colleagues or employers. In this environment a top U.S. manager may decide on a project, but implementation then requires detailed formal and written instructions and allocations of authority and responsibility among all subordinates—each of whom negotiates to protect his own individual interests. Relationships with employees, outside suppliers, and customers, which can be quite informal understandings in Japan, tend to require formalized contracts in the United States. Much time is thus required to spell out numerous contingencies and to satisfy all parties. All of this negotiating results in very slow and inefficient implementation of management decisions.

In this connection, it is noteworthy that about 500,000 lawyers practice in the United States, as compared to only 12,000 in Japan, which has about half the population. It is difficult to avoid the conclusion that most of the activities of U.S. lawyers are not really essential to society. It will be pointed out later that some of the activities of lawyers in the United States are very counterproductive, in terms of inhibiting innovation. In fact extensive reliance on formal contracts as in the U.S. seems to create—rather than avoid—disputes. Reliance on litigation as a means of resolving disputes is almost inconceivably slow and costly, compared to any other method. Lawyers clearly constitute one of the most potent special interest groups in the United States.

A further key difference between the United States and Japan is that most industry in Japan tends to be organized into hierarchical families of firms linked to major banks and trading companies. Currently there are six major groups:

- Mitsubishi (28 companies)
- Sumitomo (21 companies)
- Mitsui (23 companies, including Toyota and Toshiba)
- Fuji Bank (29 companies, including Nissan and Hitachi)
- Sanwa Bank (39 companies)
- Dai–Ichi Kangyo Bank group (45 companies)

Financing is primarily via bank debt rather than equity ownership. Banks also typically own controlling blocks of stock in their client firms. (Prior to the Bank Holding Company Act of 1933, this was the pattern in the United States.[3]) Perhaps for this reason, it is also feasible for Japanese firms to put higher priority on long-term growth and job creation in the interest of employees than on immediate profits for stockholders.[4] Whatever the explanation, Japanese firms are apparently satisfied with lower return on equity than U.S. firms would be. This adds to their ability to compete in price-sensitive markets. Moreover, the very high Japanese personal savings rate makes capital readily available for loans, while Japanese government policy has discouraged capital exports except for a brief period from December 1980 to early 1982. Banks are encouraged to lend preferentially to export industries, rather than to consumers.[5] (In summer 1982 the prime rate of interest in Japan was 9 percent as compared to 15 percent in the United States.) While the relationship between Japanese and U.S. rates varies, Japanese rates have been consistently lower during the past decade and will continue to be lower until (or unless) the Japanese again permit free capital movement. This, incidentally, results in undervaluation of the Japanese yen vis-à-vis the dollar, which in turn is a boon for Japanese exports to the United States.

There is really no reason, at least in the short term, to expect Japan's dynamic economy to stumble. Yet the problem facing the United States is not Japan itself, but the need for major structural changes in our own economic system. The United States no longer enjoys the comparative advantages in technology, manufacturing, and marketing it had in 1910. As a consequence, for most standard commodities and products, the lowest cost source in the future will undoubtedly *not* be in the United States.[6] It will shift from country to country, depending on local conditions, but generally moving away from the United States and Western Europe to Japan and the "Rim of Asia" (see Hofheinz and Calder 1982) and ultimately toward other low-wage, resource-rich countries. This movement is possible because most production technology itself is now quite standardized, easily packaged, and exportable. The older, high-wage industrial countries no longer have any comparative advantage in the mass-production industries.

WHAT A THEORY MUST EXPLAIN

This litany of economic woes has been recited not just to establish that a serious problem exists, a point most people realize already. An adequate theory must explain a great many diverse facts:

1. A declining real standard of living for most American workers (employed persons) since 1972.

2. An acceleration of the wage-price spiral since 1965, with negative real inrerest rates during most of the period from late 1972 through late 1980.

3. A sharp decline in annual per capita productivity growth, from an average of 3.2 percent per annum for 1948 to 1965 down to less than 1.5 percent per annum, on average, for the years since then.

4. A reversal in the U.S. balance of merchandise trade, from an average surplus of $4–5 billion in the 1950s to an accelerating deficit that reached $20 billion in 1980.

5. U.S. loss of share of the world's export markets for technology-intensive products from 35.5 percent in 1954 to under 20 percent in 1980.

6. Loss of U.S. domestic markets to Japanese imports in numerous industries, especially consumer electronics, watches, cameras, bicycles, motorcycles, small cars, and machine tools.

7. Deteriorating physical infrastructure, especially railways, bridges, and tunnels and water and sewer systems. Needed investments now amount to over $900 billion.

8. Aging and obsolescent productive capital equipment, especially machine tools.

9. Declining frequency of major technological innovations since the 1950s and early 1960s.

10. A deteriorating public secondary education system that produces increasingly ill educated and unprepared citizens at higher and higher cost.

A host of other symptoms could be cited, including ever-increasing litigiousness, proliferation of wasteful layers of bureaucracy (both in government and in large industry), pointless and wasteful paper entrepreneurship at the highest levels of industry, and so on. While there is some mutual causation, all of these are symptoms of some deeper social processes that we must try to understand.

The social process in question has a good deal to do with the rate of technological innovation. However, it cannot be said too often that technology change is not a prime mover or a *deus ex machina.* Technological change reflects other societal forces.

The symptoms identified seem characteristic of aging. When human beings reach and pass maturity, their bones begin to decalcify and become brittle, their joints become stiff, arteries thicken and harden, muscles lose elasticity, the rate of recovery from injuries slows down, and so on. An analogous aging phenomenon seems to hold for human institutions and nations. They become increasingly inflexible and gradually lose the ability to adapt to change. Large organizations become bureaucratic and enmeshed in rules and regulations: they do not respond quickly, either to threats or to opportunities. Even the mass production technology traditionally charateristic of large firms in mature markets tends to be highly specialized and inflexible—a significant point in terms of future technological directions.

A satisfactory explanation of all this must also explain the initiating impulses. Whether the life-cycle is that of biological organisms or societal constructs, nothing can age, however, without first being born and growing to maturity. We need to know something about the mechanisms of birth and growth.

To be sure, the biological analogy must not be pressed too far. Individuals not only grow old—they die. Rejuvenation is not possible for persons. Corporations and nations on the other hand can live on indefinitely. Senescence may, in the right circumstances, be followed by rebirth. Indeed we can see examples of this both among firms and countries. An adequate theory entails more than the elaboration of an analogy, however appealing. It must also explain how the aging phenomenon can be reconciled with generally accepted theories of the behavior of individual decisionmakers and small economic units (such as firms).

Finally, a satisfactory theory must be parsimonious, which is to say it must explain a lot with a little. There are many competing explanations for each of the trends noted in this chapter. Most of these explanations are ad hoc, treating each trend as an independent phenomenon, to be explained in terms of special character and unique circumstances. Such explanations are, by definition, not generalizable and cannot be the basis either for prediction or for policy choices. The next chapter critiques a number of the more popular explanations to lay groundwork for the discussion of technology and economic growth in Chapter 3.

NOTES TO CHAPTER 1

1. An average passenger automobile weighing 3,000 pounds or so is driven about 11,000 miles per year, whereas each heavy tractor-trailer combination weighs seven or eight times as much, accumulates around 20 times the annual mileage and probably causes 200 times as much damage to road surfaces. This is the basic justification for very sharp increases in truck registration fees scheduled to begin in 1984.

2. True, new firms were formed in record numbers in the 1970s. The high rate of bankruptcies is partly explained by this fact, since most bankruptcies occur among young firms.

3. For example, consider the role of the Mellon Bank in the development of Pittsburgh, Pennsylvania as an industrial center. The Mellon Bank played a major role in the financing of Gulf Oil Company, ALCOA, Koppers Company, Dravo, Consolidation Coal Company, and a number of other firms headquartered in the city.

4. Top executives in the U.S. firms often respond to criticisms of their focus on short-term results by asserting that Wall Street demands this. This explanation is somewhat disingenuous, but it has elements of truth. The fact that top executives tend to be financially oriented and technologically ignorant is also relevant.

5. Interest payments on consumer debt and mortgages (after the first 3 years) in Japan are not deductible from personal income taxes. This fact tends to encourage saving and discourage consumer borrowing, leaving more money for corporations and the government to borrow.

6. In a recent interview Naoki Tanaka, senior economist of the Research Institute of the National Economy (Japan), said, "As I see it, the fields in which Japan is said to be strong all have certain characteristics. Japan is by far the strongest nation in fields which involve mass production, require a high level of precision, and in which complete control of production is possible. An example would be a high-volume product such as 64K RAM (64-kilobit random access memory). In the case of custom-made products in other fields, however, the U.S. is not necessarily inferior to Japan, and Japan is not in control of all the new frontiers."

7. However much professional economists appear to disagree on macroeconomic theories, there is a fairly broad consensus on the microeconomic foundations. It now appears that Keynes's famous theory of unemployment and depression, for instance, assumes behavior on the part of individuals and firms that is not completely consistent with their rational self-interests. Non-Keynesian theories of the monetarist and "rational expectations" schools do assume rational, self-interested behavior by individuals but have not been able to explain macroeconomic phenomena such as involuntary unemployment—evidently associated with large-scale market failures that require a separate theory.

2 THE INADEQUACY OF POPULAR EXPLANATIONS

THE LIBERAL THEORIES

One popular explanation of the poor economic performance of the United States in the past few years, especially among liberal Democratic politicians, has been to blame the sharp increase in petroleum prices triggered by the "energy crisis" of 1973 and 1979–80, as exploited by the OPEC cartel. These events are widely perceived as having gravely damaged the U.S. automobile industry, which directly and (mostly) indirectly once employed nearly 20 percent of the nation's work force. The sharp rise in oil prices is often blamed for the rapidly accelerating inflation, especially after 1979. The extremely restrictive monetary policy, or tight money, which began with the arrival of Paul Volcker at the Federal Reserve Board in 1979, then forced interest rates to historic highs. Many observers attribute the deep recession of 1982 to this monetary squeeze.

Another cause of accelerating worldwide inflation in the 1970s is said to be the end of the Bretton Woods era of fixed international currency relationships and fixed gold prices. The U.S. dollar having been overvalued in the 1960s, President Nixon attempted to cure the imbalance by allowing the dollar to float. In theory, greater reliance on market mechanisms should have had beneficical effects throughout the system. But certain unexpected consequences occurred. One result was a sharp increase in the cost of imports resulting in a contribution to inflation. Much more significant was the

47

sharp rise in gold prices after 1972. This vastly increased the financial reserves and borrowing capacity of all countries with large gold hoards, including the International Monetary Fund (IMF) itself. The increased reserves soon led to increased borrowing. The higher price of gold increased the world's money supply by a large factor and thus accelerated the inflationary spiral.

These transient adjustment theories are ad hoc in the extreme. One key premise seems to be based mostly on fiction. The current troubles of the U.S. automobile industry are real enough, but most of the blame for that is probably due less to higher energy prices than to the misguided attempt by the U.S. Congress (under Democratic control) to shield American consumers from world oil prices. This policy encouraged U.S. automobile purchasers in the years 1975–1979 to spurn small cars and left the automotive industry completely unprepared without viable products when the second-round of oil price increases came in 1979–80.

As for inflation, it started to accelerate back in 1967 and reached double digit levels in 1974, dropped quite sharply in the recession of 1975, only to shoot up again to a still higher peak in 1980. Higher oil prices contributed but can hardly be given the entire blame, and probably not much of it. (Most Democrats have now abandoned this line of argument.)

CONSERVATIVE THEORIES

A collection of related explanations, shared in varying degrees (but with differing emphases) by bankers, business executives, and conservative Republicans, holds that recent economic troubles are due to a combination of underinvestment and undermotivation in the private sector and overtaxation, overspending, and overregulation by the federal government. Some also blame excessive expansion of the money supply by the Federal Reserve System.

The factual background is that the rate of fixed capital formation investment in the United States has declined: it was 30 percent lower in the 1970s than it was in 1960s per dollar of gross national product. The percentage of the GNP invested in the United States currently is less than 18 percent, as compared to 21–24 percent in Germany and France and over 30 percent in Japan.[1] In fact Japan invested 40 percent of its GNP between 1960 and 1977. Meanwhile personal savings in the United States tend to hover in the range of 5 to 7 percent of personal income, as compared to 20 percent in Japan. Investment per worker in the United States actually declined in absolute terms from 1967 to 1973. In contrast, investment per worker in Germany and Japan doubled in that period. As a result the United States has an aging industrial base: for instance, only 31 percent of machine tools were under 10 years old in 1979 in the United States, as compared to 61 percent in Japan.

It is also true that marginal tax rates have risen, pushed by inflationary "bracket creep." Government spending (federal, state, and local) in the United States rose from 27.8 percent in 1960 to 32 percent of GNP in 1979. And it is certainly true that the degree of regulation of business has grown in the 1970s. The two agencies most blamed by businessmen for excessive regulation, the Environmental Protection Agency (EPA) and the Occupational Safety and Health Agency (OSHA), did not exist at all in the 1960s. On the other hand the direct costs of compliance with environmental regulation have not been burdensome—less than 6 percent of industrial investment— has gone into pollution control equipment—and costs of compliance with OSHA regulations have been negligible. Realistically, very little of the slowdown in productivity growth can be attributed to this cause. More to the point, industries in Europe and Japan have had to pay as much (or more) for these purposes as U.S. industry. Business managers who blame the problems of U.S. basic industry on the costs of compliance with environmental regulation are simply misinformed.

The declining rate of capital formation in the United States can perhaps be traced to a long-term decline in the real rate of return on capital, the profitability of U.S. industry. Such a decline has been claimed by some economists, but questioned by others, including Fraumeni and Jorgenson (1980). An unambiguous measure of return on capital investment is fairly difficult to formulate. Inflation would partially account for such decline, if it is real. Certainly inflation has decreased the usual incentives for individuals to save for the future, since the dollars set aside have less buying power than dollars spent in the present. In other words the decline in capital formation in this country may just as well be a *consequence* as a *cause* of our economic malaise.

The undermotivation thesis—which was the basis for the massive 3-year income tax cuts of 1982–84—was based on a theory with little or no empirical justification. It is mainly predicated on a bit of speculation: the *Laffer curve*. The Laffer curve expresses the truism that, if taxes were raised indefinitely, there would *eventually* come a point where the marginal after tax return on the next increment of work is less than the value of the lost leisure time. There is little reason to suppose that effective U.S. income tax rates in the 1970s were at a counterproductive level, however. On the contrary, in 1979 only one major country in the world (Japan) collected a lower percentage of its gross national product (GNP) as taxes of all kinds than the United States.

Indeed in Germany and France in 1979 government revenues accounted for 42.9 percent and 43.4 percent of the GNP, respectively, compared to only 32.5 percent in the United States (OECD 1981). Moreover, as far back as 1960 these two countries each collected more taxes (35 percent of the GNP), than the *current* U.S. rate. Yet both Germany and France have consistently outperformed the United States in terms of economic growth and international trade throughout the two intervening decades. Deficit spending by the government, as a percentage of GNP, has also been greater in virtually every major country, including Japan, than in the United States. Most countries, including Japan, do much more income redistribution via transfer payments than the United States does. In 1977 the United States spent 14.2 percent of its GNP on welfare, compared to 17 percent for Japan, 19.8 percent for the United Kingdom, 22 percent for France, 30.6 percent for West Germany and 33.8 percent for Sweden. On the other hand the United States spent more for defense (5.5 percent of GNP) than any of the other countries.

It is also true that Europe and Japan both collect more taxes on consumption (value-added) than does the United States. The latter approach is probably one we should imitate, to provide more explicit incentives to save. However, neither of the above caveats really affects the basic argument much. The proof of the pudding is in the eating. The Reagan administration's argument for the big tax cut was that it would stimulate both private investment and private consumption, leading to an economic renaissance. Its most enthusiastic proponents even argued that it would be *self-financing* in the sense that the economic stimulus would generate more revenue than was lost by the cuts. The latter proposition has been conclusively disproven by events. The former proposition also appears questionable, to say the least. Saving and investment have not responded.

An aspect of the savings problem that has not yet received much attention outside of academic circles, is the fact that consumption or value-added taxes are inherently less discouraging to savers than are income taxes. The chairman of the Council of Economic Advisors, Martin Feldstein, has raised the issue several times. The point is that in the United States personal savings must come out of *after tax* income. Moreover, interest on savings, dividends, and capital gains on investments are taxed again as income. Dividends are, incidentally, paid out of after tax income by corporations. Thus savings are, in a sense, double-taxed in the United States. Many economists have pointed out that replacing the income tax by a consumption tax, which can be collected at the point where value is added to each product, would eliminate this problem.

It is also pertinent that the enormous complexity of U.S. tax law is due in part to legislative attempts to correct the imbalance. Thus the mortgage interest deduction is an encouragement to home ownership, which is also the major form of personal saving for middle class Americans. Similarly the mineral depletion allowance is an encouragement to invest in mineral exploration, mining, and drilling. Accelerated depreciation rules encourage certain kinds of investments. Tax-exempt municipal bonds do likewise. The vast growth of tax-deferred individual retirement accounts (IRAs) and Keogh Plan accounts provide yet another example of tax breaks on savings, to compensate for the fact that normal savings must come from after tax income.

All in all, it is likely that most personal saving *is* tax sheltered in one way or another. But the American tax system is now unreasonably complex and still discriminates against savings by people with the lowest incomes, who have little option except to put their money into passbook savings accounts. Widespread dissatisfaction with the tax system seems to be amply justified. Unfortunately, the Kemp–Roth bill did not address the most serious problems of that system.

A clinching bit of evidence against the assertion that high corporate taxes are responsible for low investment rates and poor economic performance is the fact that effective corporate tax rates in the prosperous 1950s averaged well above 50 percent, whereas since 1963 the effective rate has been below 40 percent for all except two years (1969 and 1970), as shown in Figure 2–1. But for the 1982 revision of the Kemp–Roth bill, effective corporate tax rates would have become *negative* in 1984. 1982 rates on new depreciable assets were already negative for some of the most capital-intensive industries, including motor vehicle manufacturing (–11.3 percent), mining (–3.4 percent), and transportation services including railroads and

Figure 2-1. Nominal and Effective Tax Rates, Total Nonresidential Business.

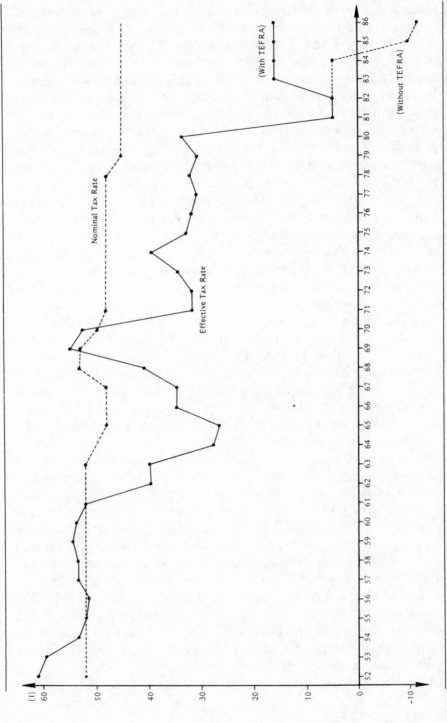

airlines (-2.9 percent). Petroleum refiners paid a modest 1.1 percent in taxes, while steel companies paid 7.5 percent. On the other hand the businesses in services and trade—many of them retail shops— were still paying corporate taxes at an average of 37.1 percent in 1982 (Economic Report of the President, February 1982: 124).

The savings rate rose slightly after the passage of Kemp–Roth, but by summer 1983 it had dropped again to a thirty-year low, as consumers resumed spending. The major initial impact of the tax cuts was to create expectations of an enormous continuing budgetary deficit that must be met by government borrowing. This has pushed real interest rates up to record high levels. From summer 1981 minimal interest rates—the *prime* rate, at which banks loan to their biggest and safest customers—declined from 20.5 percent to about 11 percent in winter 1982-83, and inflation correspondingly abated to about 4 percent for 1982 and 1983. But at the same time plant utilization was so far below capacity that investment in new plant and equipment remained virtually at a standstill.

To be sure, many leading economists from Walter Heller to Alan Greenspan and Herb Stein criticized the Kemp–Roth tax cut early in 1981 on the grounds that it would be inflationary—not that the result would be a deep recession. The recession seems to have been set off by the extremely tight monetary policies of the Federal Reserve Board (FRB) under the direction of Paul Volcker. In turn, the Federal Reserve Board was forced to respond to inflationary expectations *created* by the large federal revenue gap directly attributable to the tax cut. Perhaps a different FRB would have reacted less strongly, but in that case inflation might well have been reignited. All in all, the conservative theories are rather less ad hoc than the liberal theories but nonetheless fatally flawed.

THE "COST OF DEFENSE" THEORY

A third explanation with possible merit is that the United States has for 30 years taken major responsibility for the defense of the free world. Meanwhile other countries were able to build up their economies safely, without diverting their own resources into defense. Japan has spent very little on the mutual defense while investing very heavily in its own economic growth. Of major Western nations, the United States and the United Kingdom on the other hand have spent the largest proportion of GNP on defense over the past two decades and have the slowest growth rates. True, Germany has contributed significantly to NATO expenses and France has built its own costly *Force de Frappe*, while both countries still outperformed the United States in recent years. But it is not widely realized that both NATO allies and Japan have exacted a considerable economic price from the United States for their limited military cooperation, in the form of *co-production* agreements that amount to free transfers of valuable U.S. technology. This has helped both Europe and Japan to build up their capabilities in the electronics and aircraft industries, enabling them to compete against the United States in civilian markets. U.S. zeal to contain the expansion of Soviet Communism has cost more than we realized or admitted to ourselves.

Since 1945 the United States has undertaken to be the ultimate guarantor and military policeman of the world. At the last count, the United States maintains 359 military bases overseas, excluding bases to which we have access but do not control (*New York Times*, July 24, 1983, p. E5). We have already fought in two very expensive and nasty wars (Korea and Vietnam), and are constantly at risk of being dragged into other serious conflicts by alliance partners. A miscalculation by ourselves or by the USSR could result in a direct clash of the two superpowers, with unimaginable consequences.

Apart from the risk of nuclear or large-scale conventional war, the worldwide military-political entanglements of the United States have led to adverse economic consequences on at least four specific occasions since the Korean War. The first was the Vietnam buildup in 1967–68, when President Lyndon Johnson—perhaps suffering from the illusion of economic invulnerability—refused at first to choose between guns and butter and created an inflationary impulse in the U.S. economy. The second event was the 1972–1974 Arab oil embargo, directed at the United States because of its support of Israel in the 1973 Sinai War, which provided the excuse for a sudden and traumatic rise in the world price of petroleum. (The war did not create the circumstances that made OPEC effective, however; See the "energy-addiction" theory" in a subsequent section.) The third event resulted from the anti-Shah (and anti-United States) revolution of the Shiite Muslim fundamentalists in Iran in 1979, and the Iran–Iraq War, leading to a second sharp increase in the world price of petroleum. The fourth occasion is the Reagan administration's pursuit of its political goals of confrontation with the USSR. Reagan has insisted on raising military spending at a real rate of 7–9 percent per annum while cutting taxes. The implications for the federal budget deficit are already worrisome and projections for later years (1985–1987) are extremely adverse.

Excessive concern with maintaining and projecting military strength to influence the world has already contributed significantly to weakening the nonmilitary part of the U.S. economy. Our investment in influence through military aid has not been particularly successful either, judging by what has happened in Vietnam, Iran, Nicaragua, and Argentina, to list only four examples.[2] At the same time, ominously, we—along with the USSR—have become increasingly dependent on foreign sales of military equipment to balance our international trade budget. This gives both superpowers a vested interest, whether we like to admit it or not, in continued conflict throughout the world (as long as we do not get involved too deeply ourselves, that is).

There is much evidence to suggest that the United States has paid too dearly for "defense" since the 1950s, with little to show for it except an arms race. Nevertheless the cost-of-defense theory does not explain in any convincing way most of what has been going wrong economically for the United States since the 1960s. In fact it is somewhat contradicted by evidence that much of the U.S. technological advantage in the 1960s and 1970s resulted from military research and development (R&D) programs.

THE "TECHNOLOGY TRANSFER" AND THE "R&D SLOWDOWN" THEORIES

A fourth partial explanation for the relative economic decline of the United States is that Europe and, more so, Japan have progressed more rapidly than the United States in recent decades because they started later and could adopt our successes and avoid some of our mistakes.[3] Japan, in particular, assimilated an enormous amount of foreign, mostly U.S., technology since World War II, apparently getting it extraordinarily cheaply.[4]

There seems to be little doubt that many U.S. developments in the past have been simply copied and reproduced in other countries, notably in Japan and in the USSR with no payment. However, the Japanese rate of acquisition of foreign licenses slowed down in the

1970s, perhaps because Japan is already technologically up to date and is now a major technology innovator. In fact Japan became a net exporter of technology in 1979. Japanese penetration of U.S. markets for manufactured products has meanwhile accelerated.

The Japanese are in the enviable position of being able to wait until a new technology is developed elsewhere and then adopt and exploit it faster than the originators. This is exactly what has happened, for instance, in the case of microprocessors and dynamic random access memory (RAM) chips, which were invented by Intel Corporation in 1969 and first marketed in 1970. The so-called 1 K RAM (a unit capable of storing 1,024 words of information) was a sensationally successful product. It was produced for about 5 years entirely by U.S. based firms. By the time the next generation (4 K) RAM was ready for market in 1975, however, the Japanese were already capable of producing competitively. They got about 12 percent of a much bigger market. The 16 K RAM was introduced in 1978, and the Japanese market share rose to 40 percent of a still larger pie. Recently the 64 K RAM was introduced in 1981. The early Japanese share of this market was estimated at 70 percent. The U.S. semiconductor firms are said to be fighting back, but it will be considered a victory if they manage to hold even 50 percent of a huge market estimated at 800 million units and perhaps $20 billion per year by 1988.

A similar tale of Japanese opportunism is provided by industrial robots. Industrial robots were developed and marked in the United States as early as 1959, but U.S. customers were very slow to accept robotic technology. The first Japanese robot-licensing agreement was not signed until 1969. Some of the most advanced robot technology is still being developed in the United States, yet today Japan is far ahead of the United States in terms of both production and utilization of industrial robots, with at least 6,000 units in place by 1980 compared with 3,500 in the United States.[5] (Sweden, with 1,200 robots installed, is perhaps the most advanced country of all, on a per capita, or per unit GNP, basis.) Japanese firms have been much more willing than the U.S. firms to try out and adopt the new technology. At least 140 Japanese firms, for the most part having developed robots for internal use, are now offering industrial robots for sale. Having overcome the former U.S. technological lead, Japan now seems to be accelerating its rate of innovation in this field. In contrast, the United States is lagging in an area of technology that was formerly its exclusive preserve.

A close relative of the technology-transfer theory is the "R&D slowdown" theory. It is true that overall U.S. expenditures on R&D have declined, from a high of 3 percent of GNP in 1964 to 2.24 percent in 1979, with a very slight increase since then (Figure 2-2). The decline is entirely in the federal government contribution. On the other hand the federal cuts relative to GNP have been mostly in defense and space programs (plus some recent sharp cutbacks in non-nuclear energy research), which persuades many people that the impact on the civilian economy must have been negligible. But it has not been negligible, because of the important spillover effects of military R&D, especially in electronics during the same period. Meanwhile, R&D spending by West Germany and Japan has risen very sharply. Moreover, Japanese government R&D is mostly for basic research and economic development, areas where Japan already spends much more than the United States.

Clearly the theory depends on whether reduced federal R&D support in areas such as defense and space has had a drag effect on the civilian economy (see Chapter 3). Evidence suggests that military R&D in the 1950s was a powerful stimulus to the growth of the civilian electronics and aircraft industries, to name just two cases. It is hardly frivolous to suggest that when the level of military spending on long-range applied research in such areas was cut back (around 1969), the rate of spinoff also declined markedly. Yet these factors considered in isolation cannot account for the range of economic problems now facing the United States.

THE "ENERGY-ADDICTION" THEORY

A fifth explanation of the U.S. decline and one with considerable merit, is that the United States long ago became *addicted* to cheap natural resources, particularly petroleum. The nation's comparative advantage in manufacturing began to disappear as far back as the 1950s, as the United States became increasingly dependent on imported sources of minerals and petroleum.

The end of the long period of cheap energy in the United States occurred, rather suddenly, in 1973, when the price of petroleum im-

Figure 2-2. National Expenditures for R&D as a Percentage of GNP.

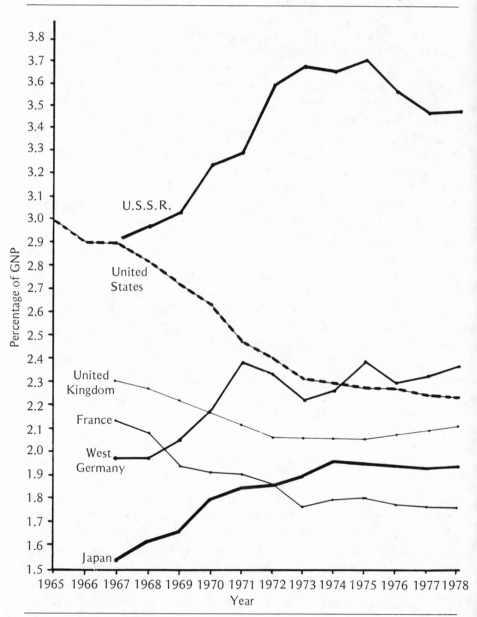

Source: OECD (1981).

ported from the OPEC countries quadrupled and prices of all other sources of energy rose sharply in sympathy. The immediate excuse was the 1973 Sinai War between Israel and Egypt, which triggered an Arab oil embargo. But the leverage that enabled the OPEC cartel to succeed in imposing such a dramatic price rise goes back to 1969, the year when U.S. domestic oil production peaked and began to decline. As a result world reserve production capacity shrank rapidly.[6]

The shrinkage of production capacity set in motion a shift from energy-intensive to energy-conserving technology, especially in the United States, that is at last beginning to gather momentum. It also marked the beginning of a serious search for new energy technologies, including technologies to utilize and convert solar energy (see Chapter 5).

The reason the energy-addiction theory has some validity as an explanation of the decline in U.S. competitiveness is that on the one hand petroleum has gradually become the lifeblood of industrialized society while on the other hand most other industrialized countries have always been forced to import most of their petroleum.[7] The fact that the United States, once self-sufficient, must now earn huge amounts of foreign exchange to pay for these imports is the most visible but the smallest part of the problem. Much more important is the fact that, for decades, all our competitors among the industrial countries have been paying much higher prices and taxes for petroleum products and other natural resources than we, as a deliberate policy to minimize foreign exchange requirements by discouraging all but essential uses. Thus Europe and Japan developed extensive electrified railway and tram systems to serve compact, well-planned cities with few sprawling suburbs or single-family homes. Small automobiles were favored not only because of high fuel taxes but also because of severe parking shortages created by strict land-use policies in Europe and Japan. European and Japanese factories and houses also were less wasteful of energy than factories and houses in the United States and Canada.

A change in the relative prices of energy vis-à-vis labor and capital altered the optimum mix and reduced the value of invested capital. From an industrial standpoint not only were most U.S. factories suddenly obsolete after 1974 but so were the products of many U.S. durable goods producers—including automobile manufacturers, although the bad news was delayed for 5 critical years by government intervention on behalf of energy consumers. The energy crisis was, in fact, a golden opportunity for German diesel-powered heavy industrial equipment, French subway cars, Japanese steel and Japanese automobiles to penetrate the U.S. domestic market, as well as to cut into remaining overseas markets for U.S. manufactured goods. The oil glut of 1982–83 provides some evidence that the U.S. has begun to adjust to higher energy prices. But temporarily lower oil prices have proved almost as troublesome as the price increases a few years ago. Unfortunately, the price decline of 1982 choked off investment in shale oil recovery, coal gasification, and R&D in solar power, especially from space. The economic problems we face are much deeper, in any case.

THE "BAD MANAGEMENT" THEORY

A sixth explanation that has recently received quite a bit of attention is the notion that American industry has been very badly managed. The newly detected flaw in the prevailing management philosophy can be described in a number of ways, but it is easiest to call it an excessive preoccupation with short-term results of existing business activities and incremental investments therein, at the expense of riskier long-term investments in new technology and market development.[8] A less sober characterization, due to Robert Reich (1983), is "paper entrepreneurialism"—a kind of "shell game" involving the rearrangement and repackaging of existing assets, rather than the creation of new ones.

The short-term preoccupation of business managers is encouraged by a doctrine that has been taught at graduate schools of business administration since the 1960s. It is that professional managers (with an MBA degree) can manage *any* business, homegrown or acquired, by analyzing financial data with a standard kit bag of methods for comparing alternative investment opportunities.[9] A large firm is thus viewed as a portfolio of businesses in various stages of maturity. The art of management consistent with this financial orientation is to balance mature businesses ("cash cows") with cash-hungry growth businesses. Business schools have largely neglected technology. They do not, as a rule, teach research management, or how to manage technological innovation. Most MBA's know little or nothing about science or engineering. Business school training provides no analytical tools for comparing alternative technological possibilities or for assessing the expectation value or option value of technology.

This combination of financial orientation and technological ignorance inevitably leads professional managers with standard business school training to undervalue technology. They tend to standardize new products prematurely, sell technology to potential competitors too cheaply, and adopt new technologies too late. They try to build conglomerates and adopt follow-the-leader strategies rather than a strategy based on homegrown technological innovation. The prevailing idea is to wait until a *growth* sector is confirmed by the marketplace. Then the conglomerates belatedly rush to acquire a piece of the action (and the technology they did not develop in house) by buying an existing company. Usually they have to pay a substantial premium over market valuation of the acquired firm. Many of the U.S. semiconductor manufacturers have been swallowed up in this manner over the past few years, in some cases with disappointing results for the acquiring firm.

Managing giant conglomerates "by the numbers" inevitably creates incentives at all levels of management to maximize current returns at the expense of medium-term growth, not to mention investment in projects needed to ensure that new products will be available to replace existing cash cows. It is plausible that conglomerates are appropriate for mature industries. But top managers with financial and accounting training have no analytic tools appropriate to measuring management performance in a dynamic (adolescent) growth sector. By applying inappropriate criteria for measuring performance, they can and often do adversely affect it.

In theory conglomerates provide compensatory savings due to greater economies of scale or scope, for example in financing and intracorporation services rendered. But in many cases both of these kinds of savings are ephemeral and difficult to realize by a dynamic performance-maximizing subsidiary.

There are a few cases on record where a conglomerate converted itself into a dynamic high-technology firm by the acquisition route. Motorola and Harris Corporation are two of the few. Much more often the merger of a small high-tech firm with a giant is an unhappy one. The previously rapid rate of growth slows down after the acquisition, and many of the key employees of the acquired firm leave within a few years. Often they proceed to set up a new, more dynamic firm in competition with their old one. Sometimes, as in Silicon Valley, the cycle of acquisitions and spinoffs is repeated many times.

A recent retrospective analysis of the ten biggest U.S. corporate acquisitions of 1971, as financial investments, revealed that the average rate of return on invested capital was actually negative in at least one case, below 5 percent in three more cases, and above 10 percent in only three cases. But not a single investment did as well over the 10-year period as the median (13.6 percent) return on capital invested by *Fortune* 500 companies (Louis 1982). Of the 1971 big 10 acquisitions, one was written off as a bad investment,[10] one was later divested, and two more were apparently for sale as of early 1982. While this particular sample of ten may conceivably be unrepresentative, there is no reason to suppose that it is anything but typical.

Given this rather dismal record, why are there so many mergers and so many conglomerates? The sad but obvious answer is that mergers and acquisitions are very profitable, both to the investment bankers who negotiate them and to top managers of the surviving firms, whose salaries and bonuses tend to rise in proportion to the size of the enterprise they control. Reich has characterized the games played by paper entrepreneurs as

> scientific management grown so extreme that it has lost all connection with the actual workplace. Its strategies involve generating profits through the clever manipulation of rules and numbers that only in theory represent real assets and products. Because paper games are always at someone else's expense, paper entrepreneurialism can be a ruthless game. It can also be fascinating and lucrative for those who play it well.

Winning a big takeover battle makes an otherwise anonymous chief executive officer or chairman of the board into a headliner—even if the long-term benefits to the stockholders are negligible or nonexistent.

Notorious in this context was the four-way takeover battle between Bendix Corporation, Martin–Marietta, United Technologies Corporation, and Allied Corporation. What happened is that Bendix offered to pay $43 per share for Marietta stock (selling for $25 at the time). The price was later raised to $48 and finally $55. Bendix ultimately bought 70 percent of the Marietta shares while the other 30 percent were withheld, at the urging of management. Meanwhile Marietta borrowed $892 million from banks to buy 51 percent of Bendix stock, also at an inflated price. If successful, Marietta proposed to split up Bendix, retaining part and selling part to United Technologies. To what purpose? Marietta thus tripled its debt, reduced its book value from $34 to $25 per share, and took on a $120 million per year added burden of interest payments, to prevent the merger. Were the stockholders protected? Or was it the top management group that benefited? Finally Allied Corporation agreed to buy Bendix stock at $80—later $85—per share, thus incidentally acquiring 39 percent of Marietta. United Technologies dropped out. Are Allied stockholders better off than before? Or was there a personal rivalry between Allied chairman Edward Hennessy, and Harry Gray, chairman of United Technologies, Hennessy's former boss, which induced Hennessey to throw caution to the winds?

In all of these shenanigans, the principals, aided and abetted by tame boards of directors, have awarded themselves lavish golden parachutes (a guaranteed payment, typically in the form of several years' salary, which becomes due in case the executive is fired by the acquiring company), worth $4.7 million per year in the Bendix case. Less than a year later all but one or two of the top Bendix officers had left or were about to do so. Several investment bankers and law firms also collected large fees, totaling perhaps $20 million from the various conglomerates. Allied's stockholders, on the other hand, have a dubious bargain, to say the least. What economic purpose was served? Probably none whatsoever.

In the words of Harvard Business School Professor Robert Hayes:

> In all these companies, workers and managers have been working hard to produce products for customers and profit for stockholders. To see these profits frittered away in such a fashion is profoundly demoralizing. Within and without, the credibility of American business management has been undermined. (*New York Times*, October 16, 1982)

American business leaders apparently see nothing wrong in this. A Harris poll for *Business Week*, conducted immediately after the affair, revealed some embarrassment. But executives overwhelmingly (89 percent) agreed that as long as everything proposed violated no laws, all of the companies involved should be able to make the moves they did in a free market economy. A majority of those polled (77 percent versus 19 percent) rejected any suggestion of congressional action to alter the existing rules. Some regulation of mergers might be desirable, even though extending the role of the federal government is not such a good idea generally. The "bad management" theory as expounded by Skinner, Hayes, Abernathy, Magaziner, and Reich explains a good deal, but by no means everything. Why, for instance, is bad management especially rife in the United States? The pernicious paper entrepreneurialism may well be only a symptom of the underlying malaise, and not its cause.

THE SOCIETAL RIGIDITY (SCLEROSIS) THEORY

A theory of group behavior set forth by Mancur Olson in *The Logic of Collective Action* (1965) was applied to macroeconomic phenomena by him in *The Rise and Fall of Nations* (1982). Several of his major theses are summarized as follows, in his own words, from the 1982 book.

- Other things being equal, the larger the number of individuals or firms that would benefit from a collective good, the smaller the share of the gains from action in the group interest that will accrue to the individual or firm that undertakes the action. Thus, in the absence of selective incentives, the incentive for group action diminishes as group size increases, so that large groups are less able to act in their common interest than small ones. (p. 31)

- Stable societies . . . tend to accumulate more collusions and organizations for collective action over time. (p. 41)

- Members of small groups have disproportionate organizational power for collective action, and this disproportion diminishes but does not disappear over time in stable societies. (p. 41)

- On balance, special interest organizations and collusions reduce efficiency and aggregate income in the societies in which they operate and make political life more divisive. (p. 47)

- The accumulation of distributional coalitions increases the complexity of regulation, the role of government, and the complexity of understandings, and changes the direction of social evolution. (p. 73)

These propositions and others that can be derived from the basic theory explain the aging phenomenon referred to earlier. The aging process consists of the accumulation of special-interest groups and coalitions or cartels in a society that has been stable for a long time. These groups exist for the purpose of redistributing existing societal assets to favor their members, even though such redistribution may result in major inefficiencies and losses to society as a whole. A major societal upheaval, by sweeping away a prior set of special-interest groups, can release the constraints and create the conditions for a burst of rapid growth.

Olson builds on these propositions to explain why the first industrial revolution occurred in Britain, why Germany grew so rapidly after its political unification, why the U.S. economy accelerated after the war between the states, and why the losers of World War II, especially Japan, have been so successful in the decades since.

Olson devotes a good deal of his attention in *Rise and Fall of Nations* to the problem of stagflation—a coincidence of high inflation and high unemployment. He points out at some length why the modern macroeconomic theories, which attribute any macroeconomic problems to temporarily erroneous expectations on the part of workers or firms, are deficient in some way or another. Although erroneous expectations, regarding inflation or deflation for example, may be able to account for some fluctuations in unemployment, no modern theory can adequately explain the high level of unemployment in Britain that persisted from the end of World War I to the outbreak of World War II, or the even higher level of unemployment in the United States from 1930 to 1940.

The older Keynesian theory is also dependent on an unexplained assumption, namely that wages and some prices are sticky and somehow inherently resistant to downward adjustments. Olson's theory successfully explains both stagflation and wage-price stickiness. In both cases the explanation rests on the formation of special-interest groups or cartels called labor unions that are organized to resist the adjustment of wages and prices toward the market-clearing equilibrium point that would maximize total output by employing all the available labor and capital productively.

Olson's theory does more. It explains why unemployment is worst among the unskilled—contrary to some modern theories—and why downward price-elasticities appear to be lowest in sectors with a few very large producers while downward wage-elasticities are lowest in the industries where labor is most organized. Moreover, on a regional basis, unemployment is highest in the states with most unionized workers as the theory implies.[11] The theory explains the interesting observation that unexpected inflation tends to result in reduced unemployment and a boom in real output (as was observed in the last Carter year, 1979–80), while unexpected deflation has the opposite consequences. Thus Olson's theory may be said to have explained the exceptionally deep recession of 1981–82, given that the credit squeeze invoked by the Federal Reserve Board was unexpectedly severe.

The Olson theory is remarkably comprehensive. It explains so much that I am almost reluctant to point out that it leaves some stones unturned. Olson himself points out that the sclerosis theory has nothing to say about the determinants of aggregate demand. In this respect it appears equally compatible with Keynes or monetarism, though they are incompatible with each other.

An equally important question is whether the sclerosis phenomenon affects the rate of technological progress—in particular, the rate of innovation. A fairly strong connection can be seen in the problem of deterioration of the nation's infrastructure. Without question, the U.S. railroads are the focus of a large number of organized special-interest groups, including a number of unions representing railroad workers, organized commuter groups, localities dependent on railway services or taxes, shippers, competitors (truckers and the Teamsters' Union), and government regulatory agencies. Amid these interested parties, railroad managements have had little room to maneuver and still less to innovate. Consequently U.S. railroad technology now lags far behind its nationalized and subsidized foreign counterparts.

Similarly, many special-interest groups have coalesced around the construction industry. Legitimate concerns for fire safety and public health, combined with concerns for architectural standards, have led to the formulation of building codes in virtually every community. Over the decades, however, these codes have come to embody standard practices. They are routinely used as protections for established construction firms and their unionized workers and as barriers to introduction of new materials and methods. Variations in code requirements between localities also make it very difficult to achieve significant economies of scale through the use of truly large-scale prefabrication techniques.

In short, though Olson does not address the application of his sclerosis theory to technological innovation, as such, it is clearly applicable to some degree. On the other hand, it is very unlikely that Olson's theory could explain the sharp worldwide decrease in technological innovation since the 1950s. Without an understanding of this core phenomenon, no satisfactory diagnosis of our economic problems is likely. Still less are we likely to find the right road to recovery. Thus the next chapter analyzes the various theories of technological change, in terms of the current situation. At the end, no simple answer will suffice.

NOTES TO CHAPTER 2

1. One reason for this difference appears to be that Japan lacks a social security system or a widespread system of pensions for retirees. Thus Japanese workers, like Americans of an earlier era, must rely on personal savings.

2. The vast investment in military aid to Israel has led to a risky involvement in Lebanon that could easily turn sour. Meanwhile U.S. influence on Israel itself seems to be minimal.

3. Thorstein Veblen theorized that Germany was able to overtake Britain in the late nineteenth century because it is easier to be a follower than a leader. But this hardly explains why Germany surpassed Britain and has continued to perform better for the better part of a century since then. It is true that they were able to adapt, buy, or borrow much technology from the United States, rather than developing it themselves from scratch.

4. From 1950 through 1981 Japan acquired over 31,000 licenses, mainly from the U.S. for a total of just over $11 billion in royalties. The Basic Oxygen Process (BOP) cost Japan only $1.2 million (Lynn 1981).

5. These figures are based on the U.S. definition of robot: a *programmable manipulator* (Ayres and Miller 1982).

6. For a good discussion of this, see Stobaugh and Yergin (1979). The exact date of the peak had been predicted (by M. King Hubbert) as early as 1954, but it still seems to have been a surprise to almost everyone in the oil industry.

7. Until the North Sea oil fields were developed in the 1970s, only the United States, Canada, and the USSR were significant petroleum producers among industrialized nations. Until recently these countries also kept internal energy prices low, in a misguided attempt to benefit their own consumers and industries.

8. It is somewhat ironic that this view is being propounded strongly, though not exclusively, by three professors at the Harvard Business School: Wickham Skinner, Robert Hayes, and William Abernathy (1980).

9. The simplest of these methods is to compute discounted present values of future cash flow and future profits. Of course, more elaborate techniques have been developed to maximize more complex financial objectives, but the spirit is the approach is similar.

10. The worst investment was General Host's acquisition of Cudahy, a meatpacker, for $80 million in cash, 141 percent of Cudahy's book value. Incidentally, $50 million would have purchased *all* of Host's outstanding stock at the 1971 high price. Nine years later Host wrote off Cudahy for $28.5 million (pretax) and has sold off the remaining assets.

11. Olson repeatedly emphasizes that unions are not the only important kind of cartel, lobby or special interest group.

3 TECHNOLOGY AS A DRIVER OF CHANGE AND ECONOMIC GROWTH

TECHNOLOGY AND HISTORICAL CAUSATION

Historians have identified many causes of societal evolution, ranging from climactic changes to tribal migrations and dynastic quarrels to religion. All the traditional explanations reflect the truth. But the most potent of all historical driving forces in the past two centuries has been technological change. Relative advantages have passed from tribe to tribe and from nation to nation since the emergence of *homo sapiens*, with discoveries in agriculture, metallurgy, navigation, weapons, and so on. Since the eighteenth century the predominance of technological innovation as an engine driving economic, social, and political change is much clearer. That economic factors superseded religion and dynastic ambition as determinants of internal and external national power, is, perhaps, the major significance of the first industrial revolution (1760–1830).

The fundamental idea that economic growth is above all a consequence of technological innovation had not occurred to most nineteenth-century economists. Karl Marx was one of the first to recognize an explicit role for technology, though he (like many economists since) seems to have regarded it as a kind of *Deus ex Machina.* In a considerable leap of imagination, Marx tried to explain social structure and evolution and class conflict as consequences of existing technology. He argued that technology made it possible for the

69

factory system run by capitalists to supersede small-scale cottage industry. His well-known view was summarized in the *Povery of Philosophy*: "The water mill gives you society with the feudal lord; the steam-mill society with the industrial capitalist." Marx was arguing that technology imposes inherent forms and constraints on social relations, especially through the work force. Thorstein Veblen, in his famous book *The Theory of the Leisure Class* (1899), also stressed the idea that work is organized to fit the requirements of machines. Neither of these writers concerned himself much with specific technologies, however. Nor did either develop further the notion of a dynamic, growing technology-driven production system. That important idea emerged only in the twentieth century, about 150 years after the beginning of the first industrial revolution.

In the 1920s and 1930s the German economist Joseph Schumpeter repeatedly emphasized the critical importance of innovation—new products, new processes—as an engine of economic growth. He wrote,

> Capitalism, is by nature a form or method of economic change and not only never is but never can be stationary. And this evolutionary character of the capitalist process is not merely due to the fact that economic life goes on in a social and natural environment which changes and by its change alters the data of economic action; this fact is important and these changes (wars, revolutions, and so on) often condition industrial change, but they are not its prime movers. Nor is this evolutionary character due to a quasi-automatic increase in population and capital or to the vagaries of monetary systems of which exactly the same thing holds true. The fundamental impulse that sets and keeps the capitalists' engine in motion comes from the new consumers' goods, the new methods of production or transportation, the new markets, the new forms of industrial organization that capitalist enterprise creates.
>
> (1950: 136–137)

Despite Schumpeter's great influence, most economists until the 1950s still believed that changes in production, or output, could be explained entirely in terms of changes in the availability of certain basic inputs: land, labor, and capital. The primacy of these so-called factors of production was axiomatic.[1] The definition of capital has always been troublesome, but most economists did not make any distinction between the *quantity* of capital (as measured in terms of original, or replacement, cost) and the *quality* of capital equipment as measured by performance capability. In recent years it has been perceived that the quality of capital, reflecting technological improvements in the production process, is more important than the quantity as measured in dollars. To cite an obvious example, a modern desktop personal computer costing $3,000 has far more capability than a 1955 mainframe computer costing around a million dollars. Other types of equipment change less drastically, but the changes are nevertheless cumulatively significant.

It was not until consistent national accounts data became available in the 1950s and older data were retroactively recompiled and revised, that economists seriously tried to test the validity of the axiomatic nineteenth century economic theory of production. If indeed output is a function of the inputs of land, labor and capital, then it ought to be possible to explain the growth in output of a country (its gross national product, or GNP) in terms of historical increases in inputs of the specified factors of production. For non-agricultural production (where land availability is clearly not a factor), the theory implies that the ratio of output to labor should be a function of capital inputs only. Similarly the ratio of output to capital should be a function of labor inputs only. These two ratios have been named *labor productivity* and *capital productivity*, respectively.

A series of empirical studies of the performance of the U.S. economy from the 1880s through the 1940s attempted to explain the increased GNP over the period in question. Indeed, three different analyses (Fabricant 1954; Abramovitz 1956; Solow 1957) were able to attribute only 10 percent to 20 percent of the increased output per man-hours (labor productivity) to increased capital investment per worker, as such. (See also Schmookler 1952, and Kendrick 1956, 1961.) Other explanatory variables were clearly needed. Solow suggested that the missing factor is "technological progress, it being understood that technological progress may be embodied in capital goods, or labor skills, or it may exist in disembodied form (as exemplified by blueprints, chemical formulae, computer programs, etc.). More recently Denison (1962, 1967, 1979) has undertaken a series of comprehensive analyses of the multiple causes of economic growth. While he found several other key change factors to be important, including demographic shifts and increasing scale of output, Denison's work elaborated but essentially confirmed Solow's conjecture on the importance of technology as a dominant driving force behind economic growth.

The complex relationship among economic growth, international technology transfer and international trade has also attracted much interest among economists since 1953. In that year Wassily Leontiev first pointed out a difficulty in the conventional Heckscher–Ohlin–Samuelson (H–O–S) equilibrium theory of international trade. According to this theory each country tends to specialize in products for which it has natural comparative advantage. Comparative advantage results from unequal natural endowments of factors of production such as land, raw materials, cheap labor, and capital. Thus countries with a high-quality, low-cost labor supply would specialize in labor-intensive products, countries with high-quality, low-cost natural resources would specialize in products made from these resources, and so on. As the factors of production were understood at the time, the endowment theory would seem to imply that a capital-rich but but high-cost-labor country such as the United States would tend to export products of capital-intensive industries. But, as Leontiev noted, the opposite was actually true. This apparent paradox prompted both a reconsideration of the nature of U.S. comparative advantage and a reconsideration of the standard theory of trade. Modifications of both have resulted.

It is now more common for economists to attribute the U.S. comparative advantage in trade not, as in earlier days, to the quantity of invested physical capital per se (except in the case of agriculture, where high-quality land and good climate are important assets), but to the *technological sophistication* of both capital and labor. In other words, if the United States has a comparative advantage in trade, it must be essentially technological in nature. This view is clearly confirmed by an examination of the recent trade performance of high versus low-technology sectors. The United States still has a strongly favorable balance of trade in high-technology projects such as computers, pharmaceuticals, and aircraft, while the balance is increasingly negative on most low-technology products, from textiles to automobiles, as noted in Chapter 1.

With regard to the standard theory of comparative advantage by virtue of static "factor endowments," most economists are still very reluctant to discard it (see Leamer 1980; see also Brecher and Choudhri 1982). But there is an increasing uneasiness over the assumption in the standard theory of a free and competitive labor market within each country. The rigidities created by institutions, notably labor unions, and the resulting nonequilibrium allocation of resources are not allowed for in the old theory. The theory fails to allow for frictional or transitional costs of adjustment, or for structural factors that may prevent the hypothetical equilibrium from existing. Basically a static equilibrium theory has little to say about dynamic nonequilibrium processes.

Meanwhile the life-cycle theory of products and industries (described later) has gained many adherents, in part because it appears to offer a simple explanation of the Leontiev paradox—namely that the United States as a high-cost country actually tends to import standard products of mature capital-intensive industries while exporting innovative products of labor-intensive adolescent industries. The latter, of course, tend to be the high-technology sectors at any given time.

WELLSPRINGS OF INNOVATION: SUPPLY-SIDE THEORIES

In most of the mainstream economics literature since Solow's famous 1956 paper, the continuing contribution of technological progress to productivity has been regarded as a kind of constant of nature—an "uncaused cause." Until comparatively recently, economists have not attempted to explore the underlying mechanisms responsible for technological innovation.

In mainstream economics supply siders are relatively scarce, and most theory has concentrated on the demand side. Not so in the case of technology, however. Perhaps this is because it is difficult to ignore the most obvious relationship in the technology system as it exists today: that modern technology generally emerges from a relatively formal and costly process of research and development (Chandler 1977; Noble 1977). In other words technological innovation may be a consequence of explicitly targetted investments in R&D, followed by even greater investments in market development.

Why should firms (or government) invest in R&D in the first place? The answer as given by Schumpeter in the quotation cited earlier is that firms in a capitalist economy compete and grow by developing new products and processes. The fundamental motivation for innovation, according to Schumpeter, is to achieve the monopoly profits that sole possession of a unique product or process conveys, at least temporarily. That a technological advantage can never be retained for long is merely a spur to further invention and innovation. Schumpeter also argued that technological innovation is primarily in the domain of very large firms, because they have the financial and marketing capability needed to exploit a new idea. This hypothesis has stimulated a vast amount of research on the relationship between innovation and firm size—unfortunately yielding few, if any, clear-cut answers.

Another question that has attracted considerable research attention for various reasons is How does investment by firms in R&D compare with other forms of investment in terms of measurable returns? On a larger scale the comparable question is Does investment in R&D foster economic growth? In other words, does investment in R&D tend to make an industry, or a nation, more competitive or faster growing than others? The U.S. Congress, through the National Science Foundation (NSF), has been an assiduous supporter of such studies, as a means of ascertaining whether government R&D dollars are well-spent (NSF 1972, 1977).

Without reviewing the various cross-sectional studies (e.g., Terleckyi, 1959, 1974) their broad conclusion is undisputed: within the United States, when different industries are compared with each other, a high level of R&D spending definitely tends to be correlated positively with other measures of economic success. Research-intensive sectors are more profitable and faster growing than other sectors. As one might suspect, however, the question of causation remains obscure. Does R&D expenditure really contribute to economic success, or vice versa? When international comparisons are made, the picture becomes muddier still—even the existence of a correlation is hard to demonstrate. Two countries with spectacular records of economic growth in technology-intensive sectors in the 1960s—Japan and Italy—invested comparatively little in R&D during that period. An obvious explanation of this anomaly is that during the period in question both Japan and Italy were acquiring advanced technology from abroad, largely via licenses, rather than from their own research laboratories. License fees, while adding to costs, were clearly not a significant obstacle to their growth.

The causal relationships have been explored from another macroeconomic perspective by Mansfield and his collaborators (1968, 1972, 1977, 1983), Mansfield has put forward the hypothesis that the rate of technological innovation is proportional to the rate of increase of the stock of R&D capital (meaning the sum of depreciated R&D expenditures). Assuming the rate of technological innovation is closely related to the rate of increased total factor productivity, the recent slowdown would seem to be explicable as a consequence of static R&D spending in the 1970s. One of their other key conclusions from numerous case studies of industrial R&D projects seems to be fairly straightforward: mission-oriented or product-oriented R&D projects in industry during the 1950s and 1960s have consistently yielded returns on investment (ROI) significantly higher than other kinds of investments.

Mansfield et al. have concluded from this that the United States underinvests in civilian R&D. This may be the case. However, an important methodological caveat must be borne in mind. Even though Mansfield's case-study approach seems to establish a direct causal link between R&D investment and the subsequent commercial exploitation, it is nevertheless possible that the specific industrial R&D projects examined were undertaken largely because exceptional commercial success could be predicted in advance. A possible alternative (partial) explanation of Mansfield's results might be that his case studies were largely taken from industries that were in a phase of rapid but orderly growth where the logical sequence of short-term product improvements and refinements could be programmed in advance by knowledgeable insiders.

Mansfield himself would probably not argue, at least on the basis of his empirical findings, that so-called basic or even long-range applied research necessarily pays off financially for the sponsoring organization. It is clear, however, that the more basic the research, the longer the interval before a potential payoff and the broader the diffusion of relevant knowledge is likely to be. Thus it is somewhat harder for a sponsor to capture the downstream benefits of basic research. On the other hand the long-run benefits of a really important technological breakthrough can be enormous—orders of magnitude larger than the benefits of a product improvement. Even though the developments cannot be totally controlled by the sponsoring organization, it is hard to dispute the fact that a competitive advantage is achieved. (The initial advantage may later be lost, but that is not a good argument for adopting a follow-the-leader strategy. The odds against followers are even greater than the odds against leaders.)

For the private sector the drawback of sponsoring long-range research is simply that it is very risky. it requires major commitments over a long period of time. Risk is somewhat easier for a public sector agency to justify, partly because the diffusion of technological knowledge to other organizations (at least within the country) can be counted as a public benefit and partly because the effective long-term discount rate (or cost of money) to a public agency is about half of that to a taxpaying private firm.[2]

The main question of causation is Does organized R&D actually foster economic growth? The available econometric evidence does not provide an unambiguous answer, but it is not inconsistent with the notion that, on the whole, R&D does pay significant economic dividends. There is really no other good explanation of the big technological lead achieved by the U.S. in computers and microelectronics since World War II. It was primarily due to spinoff from the big U.S. defense buildup following the Korean War, and continuing through most of the 1960s, that these industries got their start.

If R&D is an important element in pushing technological innovation, it follows that the educational infrastructure is of critical importance. Unless there are enough scientists, engineers, computer programmers, and computer-literate graduates of our schools and colleges in the next decade, the United States could fall behind other nations that are doing a better job of education (Botkin et al. 1982; Freeman et al. 1983). This topic is reserved for Part II, however.

WELLSPRINGS OF INNOVATION: DEMAND-SIDE THEORIES

Certainly, technology does not evolve spontaneously of its own accord. If technological evolution were an autonomous process, it would be very difficult to explain why important inventions and innovations tend to cluster in time and place as they do. The most plausible explanation for clustering is to admit the existence of underlying social and institutional forces or "demand pulls." These societal forces change over time as society itself evolves.

Sociologist William F. Ogburn was among the first to articulate clearly the notion that a societal demand of some sort rather than individual psychological or biological factors, must be the major influence in prompting new inventions and their use. Recognition of a need of some kind channels investment into the inventive process, nowadays called R&D. The existence of an external demand influences the rate of application and use of resulting inventions. Ogburn and Thomas (1922) compiled a list of 40 inventions and 108 discoveries that were made *simultaneously but independently* by two or more persons. It is hard to explain such coincidences on the basis of spontaneous generation or random genius. Ogburn and Thomas ask rhetorically, "if the various inventors had died in infancy, would not the inventions have been made and would not cultural progress have gone on without much delay?" (p. 85). The answer would seem to be in the affirmative. Several examples are noted in Chapter 4. Simultaneous or competitive invention is the rule, not the exception. Invention is clearly *not* a random process. In each case the problems of the existing technology and the need for improvements were widely comprehended and dozens of people were independently searching for solutions. History merely records the names of the most successful.

The link between societal need, or demand, and the technological response that follows was probed further by sociologist S. Colum Gilfillan (1935, 1937). Gilfillan pointed out that a need can be quite broad, so that many different approaches appear to be promising, at first. As an example that was current in the mid-1930s, Gilfillan mentioned the difficulty of navigating and flying aircraft in fog. This problem arose out of the rapid growth of commercial airlines. In other words it was actually created by an earlier wave of technological innovations. Gilfillan noted that some 25 alternative technical means of alleviating the problem could be identified, and remarked:

> With all these 25 different means apparently available for conquering fog, we may quite confidently predict that by some means or other fog will be effectively overcome for aviators soon . . . even though several of the 25 means should turn out to be worthless. . . . And hence we have a firm basis for predicting the social effects of aviation without danger from fog (1937: 22).

This quotation provides a capsule illustration of the rationale for technological forecasting. Curiously enough, the eventual solution to the fog problem—radar—was not on Gilfillan's list of 25 means. But the example merely underlines a crucial distinction, which Gilfillan was the first to make clearly: It is the societal need that generally dictates the *function* of an innovation, but of course it is the current state of science and technology that dictate the actual *means* employed and the operational details.

One mechanism that specifically signals an opportunity for individual or corporate inventors to bring forth technical solutions to societal problems is the competitive free-market price of scarce resource—inputs to the production process. Economists have considered this mechanism in relation to labor saving innovation, at least, since Hicks (1932). Hicks postulated that labor-saving innovations are motivated by historical changes in the relative price of labor (wages) as compared to capital (interest rates). An implication of this basic hypothesis of *induced innovation*, as it is now called, is that firms automatically adopt the most cost-effective labor- and capital-saving technologies that are available, thus achieving an equilibrium mix corresponding to a given ratio of wage rates (labor costs) to interest rates (capital costs). An increase in wage rates will change the equilibrium mix in favor of using marginal labor-saving techniques and not using marginal capital-saving techniques. By the same token the firms' incentive to invent new labor-saving techniques will increase. The general tendency to a more rapid increase of capital than of labor, which has marked European history during the last few centuries, has naturally provided a stimulus to labor-saving invention.

An obvious corollary of the induced-innovation hypothesis, as Hicks noted, is that a sharp rise in the price of *any* factor of production relative to the prices of capital and/or labor should call forth factor-conserving innovations. Some economists have faulted Hicks for his narrow emphasis on labor-saving innovation. Others have pointed out that innovation is only one possible response. Factor substitution is another. Nevertheless the price-inducement mechanism has apparently operated explicitly in the case of a number of major innovations over the past two centuries.

An influential study was done by Jacob Schmookler (1966), to seek empirical evidence as to whether invention causes economic growth, or the converse is true. Schmookler correlated patent applications identified with a number of industries (primarily railroading, petroleum refining and paper-making) with various indexes of economic health of those industries, over eleven decades (1840–1950). He found, that the rate of patent activity flunctuates synchronously—but on the average slightly behind—the rate of economic activity. Ergo, Schmookler's results support the hypothesis that inventive services are either directly induced by economic demand (which seems at first sight to be a denial of the "technology as driver" thesis) or that both are induced by common factors.

It would be a serious mistake to confuse the inventive services measured by Schmookler with the sort of basic inventions and major innovations of concern here, however. It is not reasonable to use the number of patent awards as a surrogate for the rate of technological innovation, as a moment's thought should make clear. The fundamental enabling inventions that made railroads possible were very few and critical: the high-pressure steam engines of Trevithick (1804), Hedley (1813), and Stephenson (1814), together with the iron rail (I or T cross-section) and the iron-wheel-on-rail configurations pioneered by Hedley, Stephenson, and Brunel, were the key items. Almost everything after 1830 was an improvement or refinement. By the early 1870s virtually every key element of the system, as it existed prior to the development of diesel-electric locomotives, had been standardized. Yet by far the majority of all railroad related patents occurred much later on. They were, of course, minor improvements, many of them never implemented. Much the same thing can be said of other industrial technologies. So Schmookler's laborious analysis of patents adds little to our qualitative understanding about the forces that actually *cause* technological innovation. It most certainly does not establish—as some have claimed—that innovation is induced by economic growth.

There are obviously other important forces at work in driving technological change. Rosenberg, (1976) emphasizes the importance of imbalances in the relationship between machines or other devices. The need for systems to permit aircraft landings and takeoffs in fog, discussed by Gilfillan, is a case in point. The imbalance resulted from rapid growth of the airline industry and lengthening of routes. Rosenberg also mentions the need for better brakes resulting from improvements in automobile engines and the need for better loudspeakers resulting from improvements in amplifiers. A thousand examples of this kind could be cited without difficulty. This is the underlying justification for Alvin Toffler's remark that "technology feeds on itself." More accurately, technological change itself helps to create the demand for innovation.

THE STALEMATE IN TECHNOLOGY

The question of most pressing importance is to discover the reasons for the apparent slowdown in the rate of innovation in the 1960s and 1970s (Figure 1–4). This slowdown has been most pronounced in the United States. The Mansfield model and the sclerosis theory of Olson provided some useful insight, but it would be well to look further.

The state of economic theory in this field is not very satisfactory. Recent attempts to address the issue start from the long-cycle hypothesis, first set forth by a Dutch economist, J. van Geldern in 1913. In 1922 Nicolai Kondratiev (1935), a Soviet Russian economist, presented empirical data showing three 50-year cycles in commodity prices, starting in 1790 and ending in 1920. He predicted another peak in 1970. (See Figure 3-1.) Kondratiev continued to work on the long-cycle hypothesis for many years, but offered no serious theoretical explanation.

In a 1939 book on business cycles, Schumpeter attempted to explain the long cycle in terms of technological revolutions, such as steam power (1818-1842), electricity (1882-1930), and the automobile (1898-1949). Schumpeter's scheme was later dissected by Simon Kuznets (1953). The cyclic scheme, as conceptualized by Kuznets, is shown in Table 3-1. Note that while Kuznets was not convinced that such a cycle exists, the intriguing pattern has continued roughly since then. The economic revival began in 1940 and was followed by a period of increasing global prosperity extending from the mid-1950s until the late 1960s, when inflation began to accelerate and growth slowed, the recession period extending to 1980 and the depression appearing to have begun in 1981. Whether this depression will continue for a decade or more (despite abortive recoveries) remains to be seen.

Schumpeter offered no clear microeconomic explanation of the dynamics of the cycle itself—though many socioeconomic phenom-

Table 3-1. Kuznets's Scheme for Kondratiev Cycles.

Prosperity	Recession	Depression	Revival
Industrial revolution wave 1787-1842: Cotton textiles, iron, steam power			
1787-1800	1801-1813	1814-1827	1828-1842
Bourgeois wave, 1842-1897: Railroadization			
1843-1857	1858-1869	1870-1884/5	1886-1897
Neomercantilist wave, 1897-1950?: Electricity, automobile			
1898-1911	1912-1924/5	1925/6-1939	1940-1953

Source: Kuznets (1953: 109).

Figure 3-1. The 50-year Kondratiev Cycle: Commodity Prices and Basic Innovations.

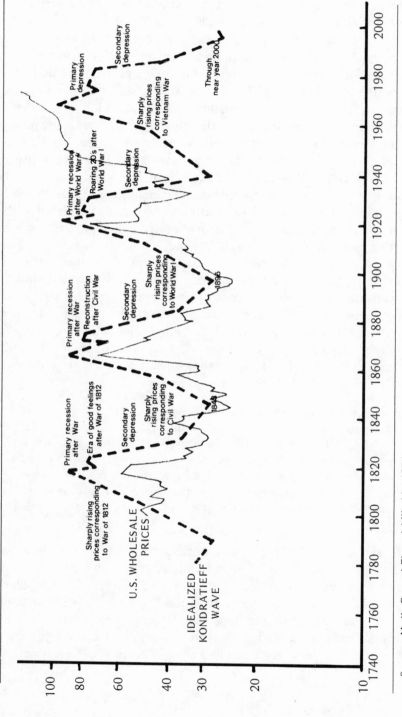

Source: Media General Financial Weekly, 1974.

ena can be roughly correlated in terms of it. Like Kuznets, many influential modern economists, such as Wassily Leontiev and Paul Samuelson, are still skeptical that a cycle exists at all.

Several mavericks from outside the economics mainstream have accepted the notion that the Kondratiev cycle is real and tried to explain it. G. Mensch (1975), Freeman, Clarke, and Soete (1982), and Jay Forrester's Systems Dynamics Group at M.I.T. (1982), all suggest that it is attributable to the clustering of major technological innovations. Mensch has assembled evidence suggesting that major innovations also occur in cycles of about 50 years, with historical peaks in 1825, 1885, and 1935. Each of Mensch's peak years coincided with an economic depression, which suggests the hypothesis that major technological innovation is inversely correlated with the level of economic activity. This is directly contrary to Schmookler's evidence about inventive activity and economic prosperity. Mensch explains this observed phenomenon in terms of the operations of the capital market: During the periods of prosperity capital flows into expansion of the most successful existing activities, as rational expectations of profits would dictate. During such periods there is little capital available for newer, riskier, innovative ventures. Eventually, however, the process goes too far: During major depressions there is severe overcapacity in the conventional sectors and capital is no longer attracted there.[3] Thus

> surges of technological basic innovations emerged after economies had fallen into a serious crisis and then passed through years of depression. . . . Since the first Industrial Revolution, whenever the economy changed from high growth to low growth in leading sectors, *the market mechanism has not been able to react quickly enough in drawing capital and labor out of the stagnating sectors and directing them into new lines of real investment* where new demand and new technologies offer new kinds of occupations. (Mensch 1975: 138)

Mensch argues at some length, anticipating Olson's sclerosis theory, that rigidities in the politicoeconomic system account for the clustering of innovations at long intervals. He blames the tax laws for attracting excess investment to capital-intensive mature industries, rather than to faster growing ones. Recent events in the U.S. economy seem to confirm his ideas in this regard. On the one hand the dynamic U.S. semiconductor industry is in need of investment capital to expand—capital that cannot be raised as equity in a depressed stock market[4] and is too expensive to borrow at inflated current interest rates. On the other hand some firms in stagnant capital-intensive industries, notably steel and gas transmission, have been awash in cash, generated from accelerated depreciation of their large fixed assets. In many cases these firms can find nothing better to do with their money then to acquire other faster growing firms in different industries. As a consequence many of the most dynamic but cash-short smaller companies have already been swallowed up by mature or senescent conglomerates, where they will be in danger of bureaucratic suffocation resulting from a tendency by top management to manage by the numbers and to impose inappropriate short-term profitability goals rather than longer term growth targets.

Freeman, Clarke, and Soete (1982) have criticized Mensch's empirical work on several grounds, one being that Mensch's list of technological innovations is subjective and that it omits a number of important process innovations. With regard to his theory of causation, Freeman argues that some of the 1930s innovations cited by Mensch were not so much responses to the depression as to rearmament pressures. Freeman et al. also argue that Mensch underrates fundamental research as a driver of innovation.

My own analysis (Chapter 1) also differs in some critical aspects from that of Mensch. The clustering phenomenon is not clearly apparent in my data, though there were major fluctuations from decade to decade that seem to deserve further study. I did not observe a major peak in the 1820s but did observe peaks in the 1860s, 1880s, 1930s and 1950s. But I do agree with Mensch's emphasis on the importance of the life-cycle model of technological evolution. In fact, so central is this concept to my own view of the problem, that a brief explanation of it is in order at this point, before I can return to the main conclusion of this chapter.

THE LIFE-CYCLE MODEL AND THE PRODUCTIVITY DILEMMA

The life-cycle concept itself is summarized briefly in Table 3–2. It was discussed in connection with products and industries as early as 1962 by Nelson (also Levitt 1965; Polli and Cook 1969; Utterback

Table 3-2. Summary of the Modified Life-cycle Theory.

Life-cycle Stage	Product Technology	Process Technology	Appropriate Competitive Strategy	Location of Production
Conception	An idea	NA	NA	NA
Birth	Prototype	NA	NA	NA
Childhood	Diversity of models and designs	Machine-specific skilled labor[a], general-purpose machines	Performance-maximizing	Near the market (in the U.S.)
Adolescence	Improved designs, fewer models, reduced rate of change	Product-specific labor skills, special adaptations of machines, e.g. tools, dies, etc.	Market-share maximizing	
Maturity	Standardized product, slow evolutionary changes	Semiskilled labor large-scale automation[b]	Factor cost-minimizing	Worldwide
Senescence	Commoditylike product		Disinvestment: sell technology, turnkey plants, management services, etc.	Mainly offshore wherever costs are lowest

a. Product-specific skills do not exist at this stage, but machine skills are especially important.
b. Automation may be "hard" or "flexible," in principle. The key to combining scale economies with continued technological change is flexible automation, discussed in more detail later in the book (Chapter 6).

and Abernathy 1975). It has been specifically applied to the problem of international trade by Vernon, (1966, 1970), Wells (1969, 1972), and Magee (1977 a, b, c). The life-cycle consists of birth and infancy; adolescence; maturity; and senescence.

Infancy begins as a new product is developed for the U.S. market—then and now the largest and richest in the world. Initial production of a new product generally takes place in the United States where the most affluent customers live. Product design in the early stages is rapidly evolving, highly variegated, and not yet standardized. It is feasible for production to be close to the market at this stage because inputs "cannot be fixed in advance with assurance" and because price-elasticity at the beginning is comparatively low, due to a high degree of product differentiation. The result is that "small cost differences count less . . . than they are likely to count later on." It is desirable for production to be close to the point of consumption to facilitate "swift and easy communication . . . with customers, suppliers and even competitors. This is a corollary of the fact that a considerable amount of uncertainty remains regarding the ultimate dimensions of the market, the efforts of rivals . . . the specifications of inputs and the specifications of the products likely to be successful." During this first stage of the life-cycle, competition in the market place is primarily based on product characteristics and performance, rather than price. (The quotations are from Vernon 1966: 195.)

The second stage, adolescence, covers the transition to large-scale domestic production. "As demand for a product expands, a certain degree of standardization usually takes place. . . . The need for flexibility declines. A commitment to some set of product standards opens up technical possibilities for achieving economies of scale, and encourages long term commitments to some given product and some fixed set of facilities. . . . Concern about production costs begins to take the place of concern about product characteristics" (Vernon 1966: 196). During this stage the basis for marketing shifts gradually from performance to price.

The third stage, maturity, is characterized by a gradual shift of new capacity to overseas production, to minimize costs. Once economies of scale in the United States are fully exploited, overseas demand can be filled more cheaply from a new facility embodying the same technology but utilizing cheaper labor. Third-country markets will be serviced from this facility, rather than from the U.S. location.

Ultimately increases in U.S. demand will also be filled from overseas plants of U.S. producers, or foreign-based manufacturers. Thus the United States ultimately becomes an importer of standardized products.

The sequence works somewhat differently for the other advanced industrial countries, and for the less developed countries, where markets develop later. Whereas the United States tends to become an importer of mature products at the end of the life-cycle, the other industrialized countries begin as net importers and later on become exporters. The Third World continues as a net importer until quite late in the cycle, but eventually, according to Vernon's hypothesis, some Third World countries such as Brazil, Taiwan, and South Korea will become exporters of standardized products to the United States. Indeed this is already happening in the case of textiles, shipbuilding, and some consumer electronics products. The sequence is illustrated in Figure 3–2.

It is implicit in this scheme that production technology tends to be relatively labor intensive early in the life-cycle, when the product is rapidly evolving and highly differentiated. Production technology will shift in the direction of capital intensity in the later stages, as the product becomes standardized. Since the initial labor-intensive stage requires co-location with the major market (production in the United States) while the final stage typically involves a shift to mass production wherever the inputs are cheapest (overseas), it seems to explain the fact first noted by Leontiev in 1953 that U.S. exports of manufactured products tend to be relatively labor intensive, as compared to U.S. imports. To cope with the implications classical trade theory has been forced to view technological sophistication as an element of comparative advantage for the United States. In ordinary language this means that for a high cost country like the United States technological sophistication is an absolute prerequisite for successful competition in trade. Unfortunately it is difficult to innovate in mature industries.

To achieve a declining long-run cost curve as an industry matures and standardizes, shown in Figure 3–3, its scale of production increases and so does the degree of automation (Figure 3–4). The more automated the process becomes, the more capital intensive it is and the longer period of time over which it must be amortized. What makes this a barrier to technological change in the product is that capital equipment specialization also tends to increase as the scale of production increases. Specialized (or so-called dedicated) high-capac-

Figure 3-2. The Product Life-cycle.

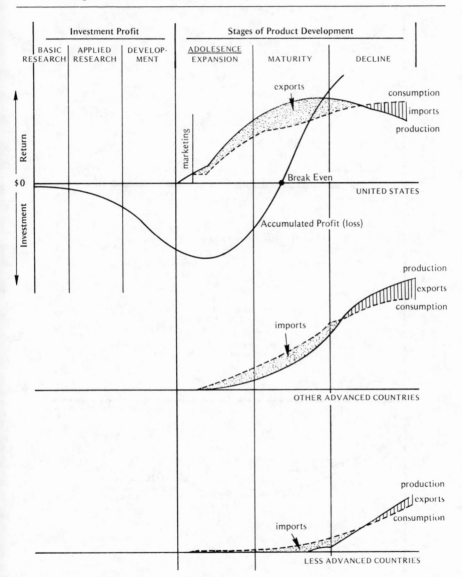

Source: Adapted from Vernon (1966).

Figure 3-3.   Unit Cost versus Production Rate.

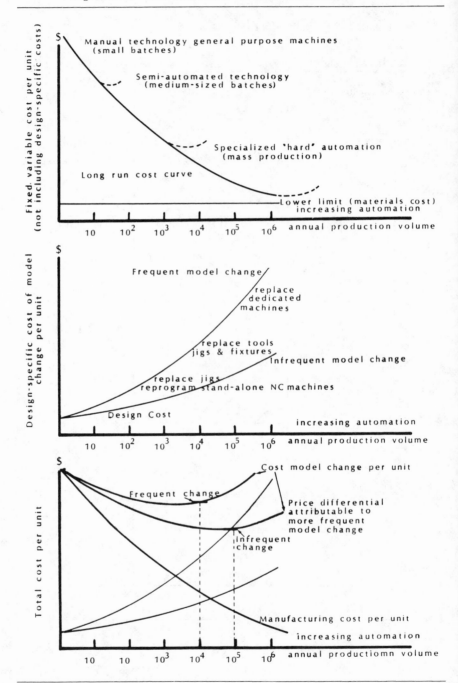

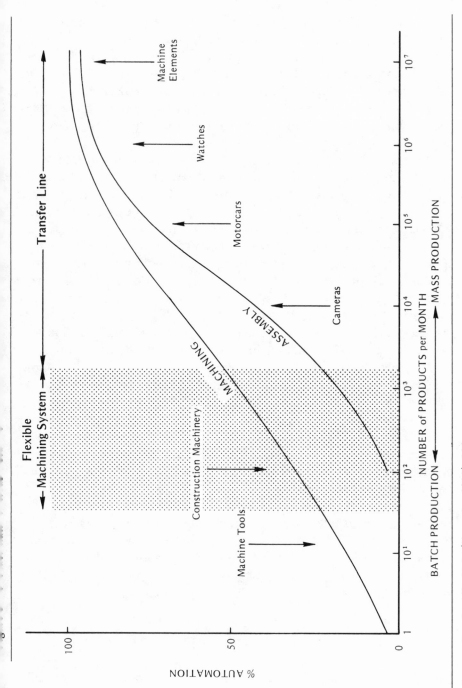

Source: *American Machinist* (November 1981): 208.

ity production lines cannot easily be adapted to produce different products. A firm with an expensive and highly specialized production facility clearly has a reduced incentive for continuing rapid technological innovation, as William Abernathy emphasized in his book *The Productivity Dilemma* (1978). Thus the transition from a growth market to a mature market is accelerated by a positive feedback mechanism. To anticipate Chapter 6, flexible automation may offer a way out of this trap.

Once a product has reached the mature stage, a major innovation is likely to be so disruptive that dominant firms based on the established technology are frequently unwilling or unable to adjust to a change initiated outside the industry. In fact the implementation costs of technological innovation grow as an industry becomes more mature, while its direct (calculable) benefits tend to decline. Abernathy (1978) estimated both the costs and the benefits for various automobile model changes before and after World War II. The average cumulative economic *value* to the innovating firm of each model change before the war was about $1.4 billion (mostly due to increased market share), the average value of model changes since the war dropped to $275 million in 1958 dollars. By contrast, the average *cost* of prewar model changes was probably less than $25 million. The Ford Mustang was introduced in 1962 at a cost of about $60 million. In 1977, Ford brought out the Zephyr and Fairmont, at a cost of $600 million, while the General Motors "J" cars, introduced in 1981 cost $5 billion, or $1 billion each for the 5 models. Given these enormous costs, the incentives to innovate are minimal, unless survival itself is at stake. No wonder innovation is often left to small fringe firms in the industry (such as Toyo Kogyo or Honda), to suppliers (such as Bendix), or to complete outsiders. Not infrequently, the large, established firms in an industry do not see the handwriting on the wall and fail to make the transition to a new technology. Historical examples are to be found among manufacturers of steam locomotives, mechanical typewriters, calculators, clocks, duplicating equipment, and so on.

EXPERIENCE, MATURITY AND SENESCENCE

The *experience curve* expresses the relationship between production cost and cumulative production. The result of numerous observations (e.g., Cochran 1968) can be summarized in the familiar empirical rule of thumb: Each doubling of cumulative production results in a fractional decline d in unit costs, where d typically lies between 0.1 and 0.4, depending on the product. In the case of a market that is growing exponentially at a constant rate, it is simple to show that costs also drop exponentially at a proportional annual rate (depending on d) as long as the market keeps on growing. In some instances, notably electronics, this behavior has continued over ten or more successive doublings of experience. Figure 3–5 illustrates the composite experience (cost) curve for electronic circuit elements.

The experience curve is now so well known that one consulting firm[5] has built a large and lucrative practice advising client firms how to base their strategic plans on building or buying market share and balancing a portfolio of businesses in different stages of the life-cycle. Under this doctrine market share is all important, since the experience curve implies that the firm with the largest market share should automatically have the lowest costs and hence the largest profits. This theory is widely taught in graduate schools of business administration, and many Japanese firms have enthusiastically adopted this American management concept.

Once a production technology is packaged and standardized, as in the case of a mature industry, the *experience* is embodied in equipment, not in labor and organization. It is thus fully transferable. Moreover, as the technology of production becomes standardized, the economics of scale dictate that specialized capital goods firms will be able to supply the equipment more efficiently than even a large producer like General Motors can do for itself. A plant embodying the latest technology can therefore be built almost anywhere in the world, given an adequate infrastructure and a literate labor force. It will normally be built, therefore, in countries where labor or energy costs are lowest.[6]

The process of technology transfer from the United States to foreign competitors and offshore export platforms has been accelerated, in recent years, by the tendency of U.S. based multinational firms to compete for market share in developing-country markets by making deals that transfer technology to foreign subsidiaries or joint-venture partners as fast as it is developed. In several key developing countries (notably Brazil, Mexico, and South Korea), entry by foreign firms to fast-growing local markets has been made conditional on supplying sophisticated technology, taking local partners (sometimes majority

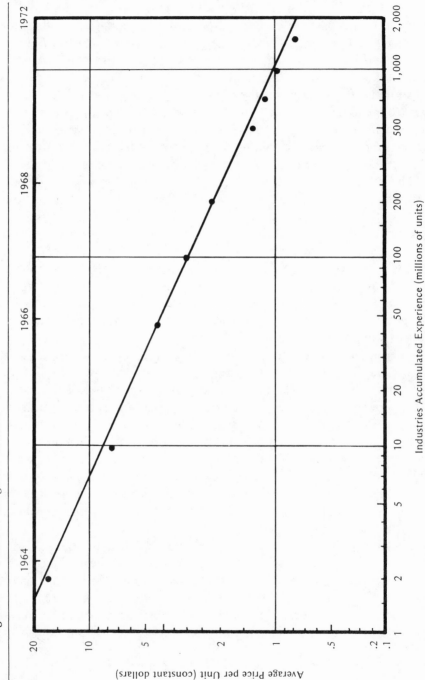

Figure 3-5. Prices of Integrated Circuits.

owners), buying from local suppliers, and manufacturing for export back to the United States (Baranson and Malmgren 1981). This development strategy has been exceedingly successful from the perspective of the more sophisticated developing countries and has presumably paid off for the multinational corporations in terms of market penetration in foreign markets.

The contribution to corporate profits made by this competition for market share in developing countries is hard to measure, but it may be presumed positive. On the other hand, if export from the United States were a viable option, the impact on the rest of the United States could well be negative, as Boretsky (1973) suggested. The adverse impact is not in doubt in cases where the product is than imported back into the United States. Jobs are forcibly shifted away from the United States to foreign countries along with the associated income and buying power, income and sales tax receipts, and foreign exchange earnings. Minor benefits to the corporate managers and stockholders are achieved at considerable real cost to the U.S. workers and communities.

As some of U.S. industry has passed beyond maturity into senescence, the process of disinvestment has become more noticeable. An example has been direct sales of "naked technology"[7] by U.S. (and other Western) firms, to Third World governments, or to nationalized firms in socialist East Bloc countries. The USSR and its satellites have been able to purchase turnkey steel mills, automobile plants, and pipelines employing the latest Western technology and often financed by below-market rate long-term government loans subsidized by Western taxpayers (in the name of job creation).[8] If the COMECON countries default on these unwise loans, it will be Western taxpayers who pick up the tab. On the other hand, if all goes well, the borrowers will repay the loans by exporting gas, steel products, and cheap automobiles back to their creditors, which will eliminate far more future jobs in the West than the turnkey projects created in the first place.

What all this means is that, unless there is a major change in the competitive environment, already standardized commodity products such as petrochemicals, basic metals, hardware, consumer appliances, many automobile and architectural parts and components, and many items of clothing will increasingly *not* be manufactured in the United States. Almost all new capacity in the future will be built in countries with cheaper resources or lower wages or both. Of the commodity products groups, only food, tobacco, glass, bricks, tiles and refractories, concrete products, pulp and paper and some wood products will continue to be manufactured on a large scale in the United States, because of the availability of local raw materials. Some of the larger existing U.S. integrated iron and steel mills may continue to produce for several more decades, being continually upgraded, but most will be gradually phased out. Probably no new greenfield steel mills (developed on a virgin site) will ever be built in the United States unless government (unwisely) subsidizes the cost. Nonintegrated (electric) minimills, utilizing steel scrap, will on the other hand continue to thrive for a long time to come.[9] Alumina and primary aluminum plants will continue to migrate to countries with good sources of bauxite or cheap electricity. Petrochemical plants will migrate to sources of cheap petroleum or gas. Instead of importing crude oil, the United States will increasingly be importing downstream products like gasoline, benzene, ethylene, and methanol. Production of radios, televisions, cameras, watches, typewriters, and a host of other standard consumer items has already migrated to Japan, Taiwan, Hong Kong and Singapore. Atari's recent decision to shift its U.S. assembly operations to Hong Kong and Taiwan is only the latest in a long series of such moves.

An adequate appreciation of the policy options and alternative futures now available to the United States cannot be gained merely from a review of what has gone wrong. The recent decline of United States' economic performance presupposes an earlier, but no less significant rise. The causes and cures of the present decline must emerge from an understanding of the causes and conditions of our former economic ascendancy.

COLLECTING THE THREADS OF THE ARGUMENT

A review of the various theories of technological change, in the light of available evidence, does not point unambiguously to a single clear answer. Neither the supply theories nor the demand theories can fully account for the phenomenon of technological progress or the recent slowdown of innovation in the United States. Some important innovations seem to be triggered by perceived *opportunities* created by technological change itself, while others seem to be triggered by specific *scarcities* of materials, energy, or labor skills. There is also a gray area in between, where either explanation might be plausible, i.e. a scarcity or "need" in some area is created by a sudden change in another. (Rosenberg uses the term "mismatch" in this situation.)

Both demand and supply-driven innovations have always been significant, but the former predominated in the first industrial revolution. However in the second half of the nineteenth century a number of important technological innovations arose, really, as a result of opportunities created by scientific discoveries in the field of electricity and magnetism.

The role of formal R&D as a causal agent has become increasingly important in the twentieth century. As Mansfield suggested, the aggregate rate of technological innovation is proportional to the rate of increase of R&D capital, defined as the accumulated sum of R&D expenditures, appropriately depreciated. This model implies that the recent slowdown in innovation in the United States is explained by the fact that R&D spending in the United States ceased to grow about 1965 and has declined in relative terms since then, as was shown in Figure 2–2.

The United States may have achieved its initial lead in the critical technologies of the next industrial revolution—computers and semiconductors—as an unintended consequence of long-range R&D programs administered primarily by the research arms of the air force and navy and a few large procurement programs at critical times, notably the Minuteman Missile system and the NORAD early warning system.

Unfortunately a controversy erupted in the U.S. research community in the late 1960s, ostensibly over the relative emphasis on undirected *basic research* vis-à-vis *mission-oriented* research. Within the U.S. Department of Defense (DoD) there were forces, emanating from both the service chiefs and the office of the assistant secretary for systems analysis, favoring the centrally planned mission-oriented approach over the less directed kinds of basic research that had long been supported. A study called Project Hindsight, carried out in 1966–67 to determine which approach was more effective, gave the nod to mission-oriented R&D. The burgeoning Vietnam conflict also contributed to pressures for short-term, quick-turnaround projects, rather than longer term research.

Coincidentally, the late 1960s was also a period of social protest, centering on the universities. One focal point of dissatisfaction was the support of a few ill-considered social science research projects by the military. Some members of the academic community also opposed classified research of any kind on university campuses. This combination of forces resulted in the so-called Mansfield Amendment to the Appropriations Act of 1969 which forbade the DoD to engage in any research not specifically tied to a military objective.

The loss of military funding of social science research, negligible to begin with, is probably of no economic importance. The cutbacks since the 1960s in general support for long-range applied research in physics, computer architecture, semiconductors, superconductors, composite materials, robotics, artificial intelligence, and so on may have significantly damaged the long-range prospects for technological innovation in the United States, however.

But the decline in undirected R&D funding cannot very well be the whole explanation of the slowdown in innovation. Two other causal factors clearly deserve note. One is that, because of a long tradition of consumerism, the U.S. government has consistently adopted policies in recent years to protect consumers from growing resource scarcities. Price controls on oil and natural gas were only one part of the pattern. More important, the United States has failed to impose *ad valorem* taxes on consumption of scarce resources such as gasoline at a rate comparable to most other industrial countries, thus depriving U.S. industry cf a major financial incentive for innovation.

A second factor of increasing importance in recent decades is the maturity of the current U.S. industrial base and its overdependence on commodities and standardized, mass-produced (commoditylike) products. An inherent contradiction between product standardization and technological innovation results from the inflexibility of specialized and dedicated capital equipment used in mass production. This problem is particularly vexing for the primary metals and automotive sectors.

The following three chapters deal with past and future technological innovations. Chapter 5 is a survey of likely future technological innovations in energy and materials. The innovations, mainly substitutions, considered in Chapter 5 are almost all in the demand or scarcity-driven category. Chapter 6 surveys likely technological innovations in the area of information technologies (telecommunications, computers, robotics). In contrast to the innovations in energy and materials, these are mostly supply (R&D) driven.

NOTES TO CHAPTER 3

1. The most quoted source seems to be Wicksteed (1894). Wicksteed is credited with the idea of the *production function*, which has since permeated economic theory. The concept was extended by Hicks in the 1930s.
2. This suggests what might be the best argument for eliminating the corporate income tax as such. The existence of the tax forces profit-making corporations to adopt a short-range perspective on investments, including R&D.
3. The Systems Dynamics Group explains the 50-year timing of the cycle in terms of the buildup of overcapacity and subsequent adjustments in the capital goods industry itself. Forrester emphasize the long lead times of major facilities such as electric power generating plants.
4. Even the recent all-time high of 1289 for the Dow-Jones average for industries is actually rather low when adjusted for inflation and compared to market levels that were reached in the 1960s.
5. Boston Consulting Group (BCG).
6. Subject to some other obvious criteria being met, notably proximity to markets, evidence of political stability, and a favorable climate for foreign investors. These conditions exist in the East Asian countries, but not in South Asia and most of the Arab world.
7. Boretsky's term. It includes licensing agreements and turnkey projects.
8. OPEC countries have also attracted capitalists to the bait. Algeria is buying a complete turnkey telecommunications system and consumer electronics industry from GTE, for cash. Saudi Arabia is buying a variety of turnkey facilities, including a complete city. Its objective is to convert its petroleum and natural gas into higher value petrochemicals, which will then be exported to the United States and Western Europe.
9. Minimills tripled their share of the U.S. market from 4 percent in 1970 to 12 percent in 1981.

4 THE FIRST TWO INDUSTRIAL REVOLUTIONS 1760-1830 and 1860-1930

An historical viewpoint is useful in the attempt to understand what drives technological change, which in turn drives economic growth. Two historical points are especially worthy of note.

1. Britain was the first country to be industrialized and the first to become rich as a consequence. It was industrial power that enabled Great Britain to defeat Napoleonic France, and to conquer and briefly hold a vast empire. *It was not the British Empire that made Britain wealthy.* On the contrary, the empire was an economic burden of incalculable but vast weight. Its costs were mostly deferred and have not yet been fully paid. Among the other costs of its empire, Britain began to lag technologically behind both Germany and the United States by the end of the nineteenth century. The British relative decline has continued since then.

2. The United States and Germany entered the twentieth century as the two strongest and most technologically dynamic industrial countries in the world. *This industrial dynamism had its roots in major technological innovations that might have taken place in Britain—but did not.* As the nineteenth century was dominated by Britain, this century has been largely dominated by Germany and the United States. Germany nearly triumphed in World War I over an alliance of Britain, France, and Russia. Czarist Russian society was shattered and the pieces were left for Lenin's Bolsheviks to reforge

into today's militaristic Communist giant. Only U.S. intervention defeated Germany in 1918—and again in World War II—and only the United States now has the power to contain the Soviet empire.

The critical question facing the present generation of Americans is this: Can the United States retain its historical technological dynamism and the economic and military power that flows from that dynamism? Or will the next generation of Americans find that the United States has been technologically and economically surpassed by Japan and the other countries of East Asia? If this happens, the Soviet Union too may escalate its ambitions. The fact that the century of Pax Britannica ended with the unutterable calamity of World War I may not be altogether coincidental. The fact that Britain finally *won* the war and Germany *lost* is almost irrelevant, as subsequent history has shown. It is doubtful whether world power could shift away from the United States to Japan or to the USSR without an even more violent period of instability, given the enormous military buildup that has been going on for decades.

THE FIRST INDUSTRIAL REVOLUTION

The first industrial revolution was centered in Britain from about 1760 to 1830. The major technologies that constituted this change were coal mining, iron making, steam power, and mechanization of the cotton spinning and weaving industry.

The societal condition that elicited these innovations was a growing shortage of wood for charcoal. Coal, available in England, Northern France, Belgium, and the Rhineland, was an obvious substitute for charcoal for many purposes. But it was not a satisfactory replacement in the smelting of pig iron or the subsequent processes used to convert pig iron to malleable bar iron. The use of coke from coal left an iron sulfide residue in the pig iron that made the metal excessively brittle. Through a combination of experiment, intuition, and luck (in his low-sulfur coal source), iron-master Abraham Darby, in the Severn Valley of western England, finally produced a coke-smelted iron with some commercial value in 1709. Darby's iron was mainly used to make high quality[1] thin-walled iron castings[1] which became the Darby specialty. Coke-smelting of iron began to displace charcoal-smelting in the 1750s and 1760s for the bar-iron market.

As a result of improvements in iron making, consumption and production of iron in Great Britain rose dramatically, from a mere
25,000 tons in 1720 to 68,000 tons in 1788 (and 1,347,000 tons another 50 years later, in 1838). Coal mining increased rapidly to meet
the demand for a cheap alternative to charcoal. Coal mining in England and Wales had already moved underground by the seventeenth
century, and the deeper mines were waterlogged. Thus another societal need was created. The only technology available for pumping in
the seventeenth century was an endless chain of buckets lifted by a
winch powered by horses or humans. Steam-powered water pumps
were designed by Denis Papin and by Thomas Savery in the last
decade of the century. A major improvement, the reciprocating
(rocker-arm) piston and cylinder arrangement, was introduced by
Thomas Newcomen in 1712 to make the steam pump cost-effective
for the first time. During the following century at least 100 large
Newcomen engines were built at various mine sites around Britain.
Although these barn-sized, slow machines averaged about 5 horsepower in continuous power output—more than the biggest wind-
mill—they achieved a thermal efficiency of only about 1 percent,
that is, roughly 1 percent of the energy value of the fuel (coal), was
converted into useful pumping work. They were thus much too expensive to use anywhere except near a source of cheap coal.

The growth of industry made possible by improved iron-making
technology created a growing demand for power to drive grinding
mills, boring mills, fulling mills for the wool industry, and bellows
(air pumps) for the iron-smelting mills and forges. Sites for water
power were growing scarce in Great Britain, which has few moun-
tains. There was a need for better stationary power sources. The solution was provided by the Scottish mechanic and inventor James
Watt and his partner Matthew Boulton. Watt made the steam engine
both smaller and more efficient than the big Newcomen engines. The
improvements came about partly because Watt had a brilliant technical idea—the separate steam condenser, patented in 1769—and
partly because by the 1770s the quality of available iron castings for
cylinders and pistons was better, due to Darby's improved casting
process.

Finally, the technology of precision boring of iron castings, to
achieve tighter fits between piston and cylinder walls, was vastly improved in 1774 by John Wilkinson, another iron-master of the Severn
Valley, mainly for the purpose of manufacturing cannons. However,

Wilkinson undertook to manufacture cylinders for Boulton and Watt's steam engines using his new boring mill. Wilkinson also purchased the first engine made by Boulton and Watt, to run a bellows for his iron-smelting furnace. Wilkinson's boring mill was, incidentally, the first of a long series of major British innovations in machine tools, associated with names like Maudslay, Roberts, Nasmyth, and Whitworth.

A further important step in iron metallurgy was to learn how to substitute coke for charcoal in the important process of converting brittle impure pig iron into malleable bar iron, which was used for virtually all iron products except castings. Key developments were the so-called "potting" process of the Wood Brothers in the 1760s and Henry Cort's "puddling-rolling" process that was perfected in the 1790s. As a result of these breakthroughs, Britain finally became the world's low-cost producer of bar iron in the early 1800s (displacing Sweden). This set the stage for the great upsurge in British manufacturing and trade over the next half century.

Metallurgical progress in turn made possible the use of high-pressure steam.[2] This was independently achieved in 1802 by a Cornishman, Richard Trevithick, and in 1804 by an American, Oliver Evans. James Watt had been afraid of using high-pressure steam, because the quality of metal available in the 1770s was inadequate. By 1800 this constraint was less severe, making higher steam pressures feasible.[3] High-pressure steam was necessary to reduce the bulk and weight of the engine. Trevithick was trying to make a steam engine powerful but compact enough to haul a train of cars on rails in an underground mine. The mine haulage system used at the time was horse-drawn carts rolling on wooden rails. Trevithick's attempt was stimulated by a sharp rise in the cost of animal feed, due to the Napoleonic wars. He did not succeed immediately because, while the engines worked well, they were still too heavy for the rails available at the time, namely, strips of iron on timbers. It took twenty years of further experimentation by many engineers to marry a steam locomotive to iron rails, culminating in George Stephenson's historic Stockton–Darlington railroad built from 1818–1825). This breakthrough opened up a new set of opportunities, setting off the railroad building orgy of the 1830s and 1840s.

Evans, on the other hand, was interested in a power source compact enough for use on boats. The first steamboat was the famous *Clermont* built in 1807 by the American Robert Fulton to operate on the Hudson River, a river too wide to bridge. Steamboats soon became the most important means of transport on the Mississippi River and its tributaries. Meanwhile steam power gradually began to supplant sails for ocean-going vessels, though sailing ships were still being built in the last decade of the nineteenth century.

The major impact of steam power, however, was on the British cotton textile industry. In the eighteenth century, Britain was primarily an importer of cotton goods from India. The domestic industry, at the time, was a cottage craft producing a rough cotton cloth for women's dresses and men's shirts. In its original form a single hand loom, operated by a skilled male weaver, was supplied with hand-spun yarn by as many as six women or children. The situation changed gradually, as a result of a series of inventions including John Kay's flying shuttle (1730), Hargreaves spinning jenny (1770), and a series of developments leading to the mechanization of spinning and weaving associated with the names of Arkwright, Crompton Cartwright, and others. In the early 1800s the first power-driven looms were being built in the Manchester area.

By 1813 there were 2,400 mostly steam-driven power looms in England, against a hundred times as many hand-looms; but by 1830 the number of power looms had grown to 85,000 in England and 15,000 in Scotland. In such a factory the productivity of a single worker was enormously greater than in the old cottage industry, resulting in a rapid decline in the price and growth in popularity of cotton products, especially fine muslins.[4] One of the saddest episodes of the first industrial revolution was the revolt of technologically unemployed cottage weavers who episodically attacked and destroyed frames and other mechanized equipment under the banner of a (probably) mythical "General Ned Ludd." The revolt was finally put down brutally with a series of trials and hangings at York in 1813. But Luddism was remembered and the word was entered our vocabulary. It is less well-known that, having finally succeeded in duplicating Indian yarns and cotton fabrics, the British in the early nineteenth century deliberately set about destroying the prosperous, competing Indian cotton-weaving industry of Bengal, eventually converting India into a captive market for factory-made British cotton goods.

But unemployment and declining income for cottage weavers and colonial abuses were not the whole story. Employment in the industry rose from less than 100,000 in 1770 to 350,000 in 1800 (Deane

1979). Prices of goods dropped sharply. Production and exports of cotton textiles soared. In 1764 England imported 4,000,000 pounds of cotton. By 1833 the figure was 300,000,000 pounds. By 1835 England produced 60 percent of all cotton goods consumed in the world—as compared to 16 percent produced in France and 7 percent in the United States.

Almost all the significant eighteenth-century inventions and innovations in iron making, steam engines, and textile manufacturing were of British origin. The only significant exceptions were Evans' high pressure steam engine mentioned above, and the Jacquard loom, associated with the French silk-weaving industry of Lyons in France. Even the Jacquard loom was adopted much more quickly in Britain than in France. Although there were many important inventions and innovations elsewhere, Britain continued to dominate in most areas of industrial technology until after the middle of the nineteenth century.

A couple of questions beg for attention at this point.

- Why did it happen in Great Britain? (Why not in France, Germany, Italy, Holland, Spain?)
- Why didn't the British hold on to their advantage and live happily ever after?

With regard to the first, there are almost as many theories as historians, but they differ in emphasis. The British defeat of the Spanish Armada in 1588 opened the world to British, Dutch, and French trade, and the English Civil War of the midseventeenth century greatly weakened royal power and strengthened Parliamentary government and the influence of the commercial classes. A long period of sociopolitical stability and an institutional clean slate following the settlement of 1688 was conducive to the growth of British trade and industry. The mercantilist policy led to an accumulation of lendable money (gold and silver) in the country and interest rates declined more or less continuously—from 10 percent in 1620 to 5 percent in 1714, to 3 percent in 1757. Not until the Napoleonic Wars was there a temporary increase in interest rates. Low interest rates were conducive to borrowing and investing in major capital projects such as turnpikes, mines, canals, and factories. During most of the eighteenth century on the other hand, economic progress in France was retarded by an overcentralized government and obsolete political and social institutions.

Other institutional factors, such as the outlawing of monopolies (1624) and other forms of vested interest in Britain also played a part (Ashton 1948). But the awarding of patents—temporary monopolies—to inventors such as Watt may have provided very useful incentives.[5] It is important not to overstate the importance of individuals. Invention is a social process, and an invention is only important if it is utilized. The spotlight of history does not rest for long on inventions that fail commercially. Many very clever inventors are merely footnotes in the history books. The successful invention fills a societal need, as a bee pollinates a flower, but it is society itself that creates the need.

Let us return to the question of why Britain did not maintain its economic and technological advantage? Note that Britain was probably passed by the United States in per capita income sometime before 1870, although GNP statistics did not exist at the time. The ruling classes of Britain never realized how far they had fallen behind until after World War I. But Britain did unqeustionably slip, and although the second industrial revolution (1876–1914) took place simultaneously in several countries, including Britain, France, Germany and the United States, the latter two countries emerged as the leaders.

No doubt the full explanation of the decline of Britain is complex. But more than anything else, the obsession with imperialism was the Achilles Heel of Great Britain. The story is worth recapitulating.

THE RISE AND FALL OF THE BRITISH EMPIRE

At the beginning of the first industrial revolution, circa 1760, Britain was, commercially and industrially, one among several more or less equal European countries. It was neither as populous nor as prosperous as France, although it had a larger colonial empire and more overseas trade. At the end of the period and indeed up to 1870 or so, Britain was the workshop of the world and by far the dominant economic power.

The British *gained* their economic ascendancy primarily through technological innovation and investment in domestic industries and infrastructure. They *lost* it largely because of a costly fascination with the Empire, although some other factors were certainly involved too. It is not particularly important why this happened. The important question for us is this: Did the imperialist era contribute on balance to British economic performance, as the imperialists themselves believed? Or did it actually absorb energies and investments that would far better have been utilized internally, as a number of contemporary skeptics, such as the socialist Hobson and the Bolskevik Lenin have argued (Winks 1963).

The latter view is most likely the correct one, though not for the reasons Hobson or Lenin gave. Prevailing opinion among intellectuals both in developed and underdeveloped countries seems to be that imperialism *must* have been good for the imperial country, since it was so bad for its colonies! There can be no doubt that colonialism *was* very bad for the colonies.[6] But it was also bad for the imperialist powers, especially the most dominant ones—first Spain and later Britain. Imperialism was not a zero-sum game.[7] The common assumption that imperialism was highly profitable is probably false. For example, Jawaharial Nehru, the first prime minister of independent India, frequently asserted that the cumulative transfer of wealth from India to Britain was in the neighborhood of 500 million pounds sterling. The origins of this popular (but dubious) idea go back to the most vociferous supporters of British imperialism themselves in the 1890s. To quote from one modern observer.

> It was a common belief among the late Victorians, if we are to go by the literature of the New Imperialism, that all these imperial activities had made their country rich: more than a belief, an assumption, for just as they did not often define their motives, so they did not generally analyze the economic situation very precisely. To the public the extension of Empire seemed more or less to have coincided with a fabulous increase in Britain's wealth, and they assumed it to be cause and effect . . .
>
> The New Imperialists represented British progress as a cycle of imperial expansion. They reasoned that the wealth of India, a century before, had provided the capital for the Industrial Revolution—which had enabled the British to acquire their new Empire elsewhere—which was itself now paying dividends." (Morris 1968: 107)

Whatever the truth may be with regard to the overall benefits and cost of imperialism, no facts appear in support of a contention that wealth from India financed the first industrial revolution. The East India Company was, of course, highly profitable to some of its individual members, who had a monopoly on export/import trade between Britain and India.[8] But in the eighteenth century this trade was never very large in relation to Britain's overall trade. The balance of the East Indian trade was actually negative for Britain: Britain imported cotton and silk yarn and piece goods from India in exchange for silver and gold, broadcloth, and various metals. In other words contrary to the prevailing mercantilist policy, the British bought goods from India in exchange for coined money (bullion), rather than the converse during the critical early decades of the first industrial revolution.[9]

True, the East India Company kept expanding its influence in India—under the British flag—and calling on the government for help whenever things got out of hand. After the 1770s the costs of expansion consistently exceeded the profits. When the British government incurred expenses in providing military assistance to the company in the mideighteenth century, Parliament insisted on repayment. In consequence the company repaid the government £400,000 a year for two years; in 1769 it agreed to continue the payments for a further five years. But in 1773 the Company was forced to reduce its dividend and borrow £1,400,000 from the government to support its growing political establishment in India. The political price of this loan was the appointment of Warren Hastings as governor general. Ten years later the Company was forced to borrow a further £900,000. As a consequence Parliament introduced further reforms in 1784 to substitute royal authority for that of the directors of the Company (Davis 1961).

Indeed according to economic historian Max Weber:

> We know that in Bengal the English garrison cost five times as much as the money value of all goods carried thither. It follows that the markets for domestic industry furnished by the colonies under the conditions of the time were relatively unimportant and that the main profit was derived from the transport business. (Weber 1950: 300)

Because of the continued deficits, the process of nationalization continued until the territories of India were fully vested in the crown in 1858, following the bloody Mutiny of 1857. The Company itself was formally dissolved two decades later.

If not from India, whence did British inventors and innovators of the eighteenth century find venture capital to back their efforts? As we noted earlier, much of the innovation in the early eighteenth century was in the iron-making industry, which grew rapidly in output and wealth. It was wealthy iron-masters such as Roebuck, Boulton, and Wilkinson who financed the development of steam power and most of the early progress in metalworking and machine tools.

Richard Arkwright, the first textile mogul, got his financial backing initially from Jedediah Strutt, inventor of a knitting machine for ribbed stockings, in exchange for needed technical assistance. Both later became independently wealthy. Strutt's initial capital was the livestock on a farm that he inherited. It looks very much as though the key innovations of the industrial revolution were essentially self-financed from profits.

With few exceptions, India being the most important, taxes collected in the colonies rarely sufficed to cover the costs of routine administration, much less of conquest or defense. Thus for the British treasury imperialism was a losing proposition from the eighteenth century on. The true cost of empire must include the costs of the military establishment needed to defend it.

From the 1840s until World War I, British foreign policy was largely determined by the Indian commitment. The Crimean War was fought to keep Russia from expanding into the Balkans at the expense of Turkey—to safeguard the approaches to India. Indeed, as one historian has noted:

> The Tsar Nicholas II once observed that all he had to do to paralyze British policy was to send a telegram mobilizing his forces in Russian Turkestan. India was to Britain like a large twin, whose hurts were felt in London as they were in Simla, and Britain's foreign policies were twisted by Indian preoccupations. As the rival Powers of Europe built up their fleets and expanded their boundaries, half the British energies were expended on securing the routes to India, safeguarding the frontiers of India, placating or overawing India's neighbors. The Trans-Siberian Railway might not seem very relevant to British prosperity, but it was a possible threat to India, and so to Britain. So was the German plan for a railway to Baghdad, and the Hejaz Railway, which the Turks were building through Arabia, and the proposed Russian railway to the Persian Gulf—and every inexplicable border dispute in the marches of Afghanistan, every French misdemeanor on the Indo-China frontier, the squabbles of sheiks in south Arabia, the weakening of Chinese power in Tibet—all, because of India, the concern of the islanders off the north-west coast of Europe. (Morris 1968: 491–492)

Not only did the British worry about the Russian threat to India. They also continued to worry about the French threat and later the German threat. The building of the Suez Canal (1854–1869) vastly increased the strategic importance of Egypt, which was ruled by the Turks. Thus, Britain, supposedly to protect Egypt, embarked on a series of ill-fated military expeditions up the Nile Valley into the

Sudan. A number of military disasters followed, at the hands of Islamic fundamentalists, all part of the cost of securing the lifeline to India. As control of the upper Nile was deemed to be necessary to maintain control of Egypt, so the territory around Lake Victoria (Buganda) came to be regarded as the strategic key to the upper Nile itself. Britain declared protectorates over Buganda in 1894 and East Africa in 1895, displacing the commercial East Africa Company.[10]

Toward the end of the century rivalry between Britain, France, and Germany over the carving up of Africa became intense. The European powers engaged in an accelerating arms race, particularly with regard to building naval ships.[11] As the Ottoman Empire continued to decay, and its abandoned Slavic provinces became targets of opportunity for the Russians—who wanted to join the imperialist game—the rivalry exploded in World War I, which paved the way to World War II. The aggregate costs of these wars in both human and monetary terms were colossal. All of the European countries were losers in the end. Germany and Russia were physically devastated. Russia has suffered under the yoke of Communist tyranny since 1917. For the French the costs continued into the 1950s with the agony of the Indo-Chinese and Algerian wars of liberation. The British paid less dearly than the French to dispose of their colonial possessions, but since World War II the costs have continued to mount—in Malaysia, Palestine, Aden, Cyprus, and, most recently the Falkland Islands.

The late nineteenth-century British obsession with projecting military power throughout the world to protect its lifelines suggests an ominous parallel with the United States foreign policy today.

There were, of course, other reasons for Britain's decline from economic preeminence a century ago to is present secondary status. A list of the reasons that have been cited would include the following:

1. The ruling class of Britain was drawn from the land-holding aristocrats and landed gentry. Primogeniture meant that younger sons could not inherit estates in Britain. This created strong pressures to open up new lands (under the British flag) for planters, administrators, and military men. The land-oriented governing class was on the other hand relatively uninterested in domestic industrial development.

2. The social system of Britain became rigid and created severe strains. Entrepreneurs from lower orders needed sponsorship

from nobility—a lord on the board. The upper class looked down on, and patronized, anyone in trade. By contrast, successful entrepreneurs in the United States were admired and treated as a kind of aristocracy.

3. Other countries had greater (perceived) needs.

4. Other countries learned to adapt technology initially developed in Britain (e.g. aniline dyes).

5. Other countries (Germany and the United States) did better at educating their work forces.

Of these, the persistence of social distinctions based on class origin may be the most important. There seems little doubt, for instance, that labor-management relations in the United Kingdom are consistently worse than in most of the rest of the industrialized world and that this antagonism is due in part to the long-entrenched class system. (But poor labor relations may well be attributable in part to slow economic growth.)

Another factor that will be discussed subsequently was the lack of stimulus to find substitutes for raw materials obtained from the empire. Vested interest in colonial agriculture and mining ventures inhibited investment in technological alternatives. Imperial Britain, with its numerous tropical colonies, was not greatly interested in synthetic nitrates, synthetic dyes, synthetic rubber, synthetic fibers, and so on. Britain was effectively deprived of the incentive of resource scarcity that drove Germany, its chief economic rival, to build a strong technological base in several important fields.

THE SECOND INDUSTRIAL REVOLUTION: 1860–1930

The second industrial revolution, which began in the 1860s, was in many ways bigger than the first, which was mainly focussed on iron-making, steam power, and the cotton textile industry. It affected more industries, certainly, and had a more immediate impact on ordinary people.

The second technological revolution comprised four major areas of technological innovation and a host of minor ones. The major ones were the shift from iron to steel as an engineering material, the practical application of electricity, and the internal combustion engine (a prerequisite of the automobile and the airplane), and mass production of consumer goods.[12]

The earliest of these innovations was steel manufacturing. By the midnineteenth century the spread of steamships and railroads—driven by more powerful locomotives at higher speeds—created an acute need for better engineering materials. Wrought and cast iron were no longer hard enough or strong enough. Industry needed steel. The problem was to make it in large quantities but without sacrificing quality. The cluster of key developments in steel-making have been called "almost the greatest invention" (Morrison 1966), because of their vast ramifications throughout society.

To make steel in large quantities required a method of rapid removal of unwanted carbon and other impurities from molten pig iron. The problem was simple: pig iron has a much lower melting point than pure iron or steel, precisely because of its carbon content. Thus as pig iron is purified it tends to harden—thus inhibiting further reactions. Conventional blast furnaces using charcoal or coke could achieve a high enough temperature to melt pig iron (1150°C), but not a high enough temperature to melt steel (1500°C). Thus all processes to purify pig iron prior to the 1860s involved at least one stage of removal of impurities from the surface of a solid or semisolid lump of metal. Obviously such a process could not be efficient if the lump of metal was too large, because the surface-to-volume ratio decreases with lump size. Hence the quantities that could be purified at one time were necessarily small.

The trick that finally succeeded in avoiding this difficulty was developed simultaneously over a period of years, by Henry Bessemer in Britain (1855-60) and William Kelly in the United States (1847-56). It was, essentially, to blow air rapidly through a large container of molten pig iron. The air oxidized the carbon, thus purifying the iron, and also produced enough heat to keep it molten. By this method many tons of pig iron could be "converted" in a few minutes. The final ingredient, necessary to produce good quality steel, was an additive called "spiegeleisen," the use of which was perfected by Robert Mushet from 1856 to 1859.[13] By 1860 the Bessemer process was ready for wide use.

For all its importance, the Bessemer process was not completely satisfactory. We now know the reason: because dissolved nitrogen (from the air) sometimes made the steel too brittle. A completely different approach, pioneered by Emile and Pierre Martin of France (1863), led to the familiar open hearth furnace, using a steam regenerative heating[14] system developed by W. Siemens and T. Cowper in 1856–57. The use of gas as a fuel, plus regenerators to conserve heat, made it possible to achieve the high temperatures necessary to keep the metal molten as the excess carbon is oxidized. By 1890 most of the new steel furnaces used this process—despite its relative slowness—because of the higher quality of the product. In 1867, only 2,500 tons of Bessemer steel rail were made in the United States, at $170/ton—compared to 460,000 tons of iron rails, at $83/ton. As steel output rose, prices dropped sharply. By 1884, the last year of iron rail manufacture, production of steel rail was up to 1,500,000 tons and prices were down to $32/ton. The price of steel dropped to a low point of $15/ton in 1898, when total output reached 10,000,000 tons. After a century of nearly constant price for bar iron, a much better material became available at only a fraction of the price. This revolutionary change occurred in about thirty years. The key technologies originated in Britain, the United States, and France.

Early discoveries in the field of electricity and magnetism were shared between England, France, Italy, and the United States. Early contributions were due to Volta (the voltaic pile), Ampére, and others. Joseph Henry, an American, was first to discover that electric current is induced by a changing magnetic field. This discovery was repeated independently by Michael Faraday at the Royal Institution around 1820. Faraday invented a crude form of electric motor in 1821. Scientific work on electricity and magnetism went forward in several countries, culminating in the great theoretical work of James Clerk Maxwell, who—among other things—predicted the existence of electromagnetic waves and explained the electromagnetic nature of visible light. The wide range of potential opportunities created by these discoveries was recognized early and stimulated much research and invention.

But major practical applications of electromagnetism were not immediate. The electric telegraph (1840s) was the first of them, followed by the arc light (1850s). The dynamo (generator) appeared in the 1860s (Nollet and Holmes; Siemens and Halske), the DC motor appeared in the 1870s (Gramme) and the incandescent light arrived in the 1880s. Commercial-minded inventors like Samuel F.B. Morse,

Werner Siemens (brother of Sir William Siemens), Alexander Graham Bell, Thomas A. Edison, Frank Sprague, Charles Brush, Elihu Thomson, and George Westinghouse spawned great industrial firms in the late nineteenth century in Germany and the United States. From Edison's first commercial electric generating plant in 1882, electrification of cities and factories proceeded rather slowly for a decade, mainly because of technical disputes over the superiority of direct current (DC) versus alternating current (AC). Electric street railways and lights were the first major applications.

After 1895, as practical long-distance power distribution systems were developed by German and Swiss firms, and by Westinghouse in the United States, progress in the electrification of industry began to accelerate. Niagara Falls was tamed for hydroelectric power in that year, creating a reliable low-cost supply for the northeastern United States. Other hydroelectric sources were soon tapped. In 1899 there were 16,891 industrial electric motors in the United States, having a total capacity of under 500,000 horsepower. But 10 years later the number of industrial motors had grown twentyfold to 388,854 and installed capacity had risen tenfold, to 4,817,000 horsepower. By 1929, electric motors had almost totally replaced steam engines, water power, or other prime movers in factories. Meanwhile the cost of electricity was dropping rapidly as generating plants grew larger and increased in efficiency. By 1910 the major cities of Western Europe and the United States were electrified, and many urban homes had electric lights, not to mention other electric appliances such as electric sewing machines (Singer), electric vacuum sweepers (Hoover), and electric irons. However, the period of most rapid penetration of electric appliances, such as washing machines, kitchen ranges, water heaters and refrigerators into the average home took place mostly in the 1920s.

The great industrial enterprises spawned by the applications of electricity have been preeminently American and German-Swiss in origin. It is hard to fathom why no major firm in this field developed in the United Kingdom, despite the important key scientific and technical contributions of Britons. But such is the fact.

The third major area of technological innovation in the 1870–1930 period was the internal combustion (gas) engine, which ultimately made possible the automobile, truck, tractor, motorcycle, and airplane. Again the early developments were mostly British. But after 1860 the center of technological activity moved, for reasons that are unclear, to France and Germany. In 1860 Etienne Lenoir (France) built the first commercially successful two-stroke double-acting atmospheric[15] internal combustion engine, using gas as a fuel. Lenoir's engine was soon copied and then improved upon by Nicolaus Otto in Germany. With a partner, Otto began production of small stationary gas engines, used as prime movers in factories in lieu of steam engines. From 1865 to 1876 the firm of Otto and Langen produced and sold about 5,000 such engines. Then came a big breakthrough that dramatically increased power and efficiency while reducing weight. The so-called "silent Otto" of 1876 produced an amazing 3 horsepower at 180 rpm, with a very compact design, by compressing the fuel-air mixture prior to ignition.

It was this improvement in power output per unit weight of engine that finally made the self-propelled vehicle a feasible objective. Gottlieb Daimler and Wilhelm Maybach, young engineers associated with Otto, became interested in applications of the high-pressure internal combustion engine to the requirements of a vehicle. They joined forces with Karl Benz, an independent inventor and entrepreneur with similar interests, in the early 1880s. Thus emerged the firm of Daimler–Benz, the first and oldest surviving automobile manufacturer and the world's largest producer of buses and trucks.

The advantages of high pressure particularly impressed the young engineer, Rudolph Diesel. He realized that ultra-high-pressure combustion could result in extraordinary economy of operation, as well as eliminate the need for spark ignition. Rudolph Diesel's engineering efforts, beginning in the 1890s, led to the application of diesel engines to ships soon after 1900, and to railroad locomotives around 1912. Engines compact enough for trucks did not finally become practical until after 1920. Germany has continued from that time to lead the world in diesel engine technology. The challenge of solving a host of problems associated with the design and construction of compression-ignition ultra-high-pressure engines has left a valuable engineering legacy that probably accounts for much of Germany's continuing strength in heavy-equipment manufacturing industries generally.

Why did Germany take the lead in this new field of technology? The answer may be in part that Germany made a later and lesser commitment than Britain to steam power. Moreover, Germany was industrially dynamic but land and energy poor, and needed to use its only existing energy resource (coal) efficiently.[16] Gas as a by-product of coke ovens was readily available in the Ruhr area, in particular. The development of mobile power sources, of course, required liquid fuel, which had to be obtained from imported petroleum, mostly from the United States until 1900 when the Caucasian fields became an alternative source of supply. Middle East production of oil began in Iraq and Kuwait in 1908, but this source was dominated from the start by British and American firms—backed up by their respective navies. Middle Eastern sources were thus not reliably accessible to Germany. These factors certainly provided a major incentive for development of the most efficient possible internal combustion engine and synthetic fuels in Germany. But high-quality engineering education and a strong program of industrial research, encouraged by the government, were also important.

The fourth and distinctively American contribution to the second industrial revolution was mass production, associated in the public mind with Henry Ford and the automobile industry, which began to grow rapidly in the early 1900s. Ford's famous ideas about the organization of production had peculiarly American antecedents, going back at least to the ingenious Connecticut Yankee inventor, Eli Whitney. Whitney's cotton gin (1794), to separate cotton seeds from the lint, was the first great labor-saving invention in the United States. Whitney soon built a factory to produce cotton gins near New Haven, Connecticut. He sold as many as he could produce but was plagued by unlicensed imitators. By 1798 he had also began the production of 10,000 muskets under contract to the U.S. army. It took him over 20 years to complete the order in small batches, and the effort incidentally provided a great stimulus to the New England machine tool industry.

Whitney's basic problem, as a manufacturer both of gins and guns, was inability to meet the demand. There were simply not enough skilled metalworkers, and he was under constant pressure to find ways of simplifying the production process by mechanization, division of labor, and by standardization of parts. This particular societal need was unique to the United States. European countries were poor in materials resources but had ample skilled labor. Whitney deserves considerable credit, in particular, for refining the concept of interchangeable parts, although he was never able to implement this notion completely with the machine tools available to him (Woodbury 1960). In fact Whitney's greatest invention—a product of his need for better machine tools—may have been the milling machine (1818).

The ideas of Whitney were considerably advanced by another Connecticut Yankee, Samuel Colt, inventor of a semiautomatic pistol with a many-chambered rotating breech—the so-called revolver. After the Mexican War resulted in a sharp increase in demand, for the pistols, Colt built new factories incorporating the use of fully interchangeable parts—made possible by more advanced machine tools—and a production line similar to the modern assembly line. Colt exhibited both his revolvers and his methods of production at the epochal Crystal Palace Exhibition in London (1851), creating something of a sensation (Rosenberg 1974) and even became the subject of a Parliamentary investigation. Indeed the Colt manufacturing system—which was soon imitated by other New England entrepreneurs—became known in Europe as the American system of manufacturing. Though England was still known as the workshop of the world in 1850, it no longer boasted the most sophisticated manufacturing technology.[17]

The mechanization of industry, barely begun by 1850, was far advanced by the 1890s. Increases in output per man-hour ranged from five-fold to fifty-fold or more. Table 4–1, extracted from a massive contemporary study by the U.S. Department of Labor (1895) underlines this point. This radical change created the industrial capability required for true mass production, as exemplified by Ford.

The trend toward rationalization of the manufacturing process was advanced from another direction by Frederick Winslow Taylor. Having started as an apprentice machinist and pattern-maker, Taylor became a gang boss for Midvale Steel Company (1880), where labor problems impressed on him the value of being able to define precisely the time required for each operation. Taylor subsequently undertook many intensive time-and-motion studies of specific tasks, became a well-known consulting engineer (1893–1901), and wrote several path-breaking books on scientific factory management. Taylor's work certainly influenced Henry Ford.

By 1895 the United States still lagged several years behind European automotive technology, perhaps because roads were poor and distances were so great. But a growing, prosperous, and dispersed population needed more flexible means of transportation than rail-

Table 4-1. Productivity Increase Due to Mechanization.

Item	Period	Increased Output per Man-hour (Multiplier)
Metal Products		
pitchforks (steel)	1836-1896	15.6
plows, iron and wood	1836-1896	3.15
rakes, steel	1858-1896	5.96
axle nuts (2")	1850-1895	148
carriage axles	1856-1896	6.23
carriage axles (4" steel)	1862-1896	6.23
tire bolts (1 3/4" × 3/16")	1856-1896	46.9
carriage wheels (3'6")	1860-1895	8.41
clocks, 8-day brass	1850-1896	8.30
watch movements, brass	1850-1896	35.5
shears, 8"	1854-1895	5.51
saw files, 4" tapered	1872-1895	5.51
rifle barrels, 34 1/2"	1856-1896	26.2
welded iron pipe, 4"	1835-1895	17.6
nails, horseshoe, no. 7	1864-1896	23.8
sewing machine needles	1844-1895	6.7
Other Products		
bookbinding, cloth (320 pp)	1862-1895	3.80
mens shoes, cheap	1859-1895	932
womans shoes, cheap	1858-1895	12.8
hat boxes, paperboard	1860-1896	3.22
wood boxes (18" × 16" × 9")	1860-1896	9.73
paving bricks	1830-1896	3.89
buttons, bone	1842-1895	4.04
carpet, Brussels	1850-1895	7.95
overalls, mens	1870-1895	10.1
rope, hemp	1870-1895	9.74
sheet, cotton	1860-1896	106
electrotype plates	1865-1895	2.91
chairs, maple	1845-1897	6.43

roads could provide. Manufacturing capabilities were unsurpassed. Petroleum was being discovered in vast quantities in Texas, and liquid fuel was rapidly becoming cheaper. Conditions were ideal for production and sales of automobiles on an enormous scale. The Census of 1900 reported 4,292 horseless carriages wre produced in the United States. The biggest manufacturer in that year was Stanley Motors of Portland, Maine. Average price of a car was $1,000, and vehicle reliability was poor. By 1908, the year of the introduction of the Ford Model T (and the founding of General Motors) production rose to 65,000 units and total automobile registrations to 400,000. There were scores of small manufacturers, but Henry Ford had already begun to change the structure of the industry. His contribution to technology per se was minor. His chief innovation was the moving assembly line (1918) and mass production, combining the principles of division of labor, mechanization of processes, interchangeability of parts, and rational scientific management. Ford's breakthrough created a vast new economic opportunity, which he and others went on to exploit vigorously.

From 1910 to 1920 U.S. automobile output rose tenfold while prices declined by 62 percent in real terms. (When the Model T was finally discontinued in 1926 its price was down to $300.) Yet employment in the industry increased from 37,000 to 206,000 in this 10-year period. Automobiles became cheap enough and reliable enough to be practical personal transportation for virtually anyone. The age of mass production (and marketing) had arrived. The United States became—and remains to this day—the exemplar of the consumer society (see Williamson 1967).

THE ROLE OF ORGANIZED RESEARCH

A notable feature of technological change in the eighteenth and most of the nineteenth centuries was the casual ad hoc nature of most invention. The formal link between scientific discovery and technological innovation was almost nonexistent until Napoleon Bonaparte created the Ecole Polytechnique to foster technological progress. However, Napoleon's innovation was soon imitated and improved upon by the Germans.

From the founding of Justus Leibig's famous laboratory at Geissen in 1825, Germany pioneered in organized applied research supported by government grants. In addition, Germany invested in a high-quality public school and university system. What drove Germany to do these things was lack of land and natural resources. One of Leibig's major concerns, for example, was how to make his nation more nearly self-sufficient in food. The addition of mineral fertilizers to supplement needed elements in the soil began in Germany. The long quest to learn how to fix atmospheric nitrogen to replace depleted soil nitrates culminated, at last, in the famous Haber–Bosch process to synthesize ammonia (1914). The quest itself strengthened the German chemical industry immeasurably. Leibig also began a systematic study of hydrocarbon chemistry, which enabled Germany to take advantage of the availability of coal tar as a source of useful products, including synthetic dyes. The innovativeness of the German chemical industry enabled Germany to produce synthetic rubber and gasoline from coal in World War II, despite almost a total lack of petroleum.

The first organized center for R&D established in the United States promote applied science and the mechanical arts was the Franklin Institute, founded in 1824 in Philadelphia. The work of the Institute played a key role in the development of machines and machine-building industries in the United States in the early nineteenth century. Organized industrial research in the United States has obscure beginnings but flowered in the electrical industry (Noble 1977). Edison's Menlo Park, New Jersey laboratory may have been the first major organization devoted to mission-oriented R&D on demand. The experimental department of the newly formed Bell Telephone Company was created in 1879 by the first general manager of the firm, Theodore N. Vail, with the explicit purpose of controlling all possible inventions relevant to the telephone business. This was the origin of the Bell Telephone Laboratories. In the last two decades of the nineteenth century major industrial laboratories were also created by several other firms, including General Electric Corporation and Westinghouse (Noble 1977).

The United States began to build a formal system to support applied research in agriculture and the mechanical arts beginning with the Land Grant Act of 1962 (first Morrill Act). The federal government initially endowed the system with 17,430,000 acres of land to subsidize the founding of institutions. A second Morrill Act in 1890 extended the system to states. There is much evidence to suggest tht current U.S. proficiency in agriculture is largely due to the creation of this system of educational and research institutions.

In comparison with France, Germany, and the United States, Britain was slow to develop government supported institutions for basic or applied research.

CONCLUSIONS

An interesting question is Why was the United States the major beneficiary of the second industrial revolution, especially in its later stages? A related question is What role was played by the availability or lack of resources?

As to the first question, many explanations have been put forward. The one usually cited by economists is the availability of a large and rapidly growing market. U.S. population more than doubled from 40 million in 1870 to 92 million in 1910 and per capita income more than doubled. Total national output (GNP) quadrupled in that period, making the U.S. economy by far larger than any other in the world. Clearly this is an ideal situation for any manufacturer, especially one with a hot new consumer product to sell.

The United States also had a well-educated[18] and highly mobile work force that adapted well to factory conditions.[19] Certainly the United States benefited at that time from the availability of cheap domestic supplies of industrial raw materials, especially petroleum. As noted, factory mechanization was far advanced in the United States. These factors, plus economies of scale, enabled manufacturers to pay fairly high wages[20] while still competing quite successfully on world markets.

The second question recalls the generalized Hicks induced-innovation hypothesis mentioned in the last chapter, namely that unavailability of a critical (nonsubstitutable) resource can be a very effective inducement to innovation. In fact any event that serves to create apprehension in regard to future availability of the resource may suffice—even in the absence of any immediate price increases.

Labor, in the aggregate, is a nonsubstitutable (critical) resource. While any particular worker can usually be replaced, the simultaneous withdrawal of the whole work force (by a strike) means that production must cease. Thus the threat of strikes has been a potent inducement on occasion to technological innovation. For instance, the invention of the self-acting (spinning) mule by Richard Roberts

(1825) was in direct response to a strike by the Manchester mule spinners. Other examples included the rivetting machine, invented by W. Fairbairn (1837), Nasmyth's steam-powered forging hammer (1839) and other self-acting machines that were invented in Britain (Rosenberg 1976).

By contrast, American developments in this area were primarily stimulated by labor shortage rather than strike threats. Major nineteenth century innovations in the United States were concentrated primarily in three areas:

1. Agricultural mechanization
2. The American system of manufacturing based on the use of interchangeable parts to eliminate fitting
3. Labor-saving consumer products

Hired farm labor was unavailable, in practice, because land in the United States was so cheap in the period 1800–1850 that any able-bodied man with appropriate skills could buy an acre of land for himself for as little as a single week's wage. In comparison, in England, an acre of farmland cost as much as a year's wages (Christiansen 1979). Obviously in the northern United States the only choice for a farmer was to breed as many sons as possible and—whenever possible—to mechanize.

Similarly, skilled mechanics (machinists) and metal workers were comparatively scarce in the United States, whereas and partly because the market for metal products such as hand tools, farm implements, guns, wagon wheels and axles, not to mention clocks, was burgeoning. Consequently there was a societal need for more efficient, less skilled, labor-intensive methods of manufacturing, particularly of metal products (see Rosenberg 1972: 87–90).

Finally, there was a shortage of household servants in the northern states. Any middle class family in Europe could and did have several live-in female servants, daughters of local agricultural laborers and peasants, to undertake labor-intensive chores such as food processing and preserving, sewing, laundry, and cleaning. In the United States such servants were much scarcer and more expensive, because of the lack of peasantry, and the middle class small town or rural housewife readily utilized mechanical assistance of all kinds. This led to a host of labor-saving inventions from apple-corers and ruffleironers to the sewing machine, washing machine, and vacuum cleaner. Indeed the Model T was essentially a utility vehicle for the rural middle class.

Material and energy scarcities had a greater impact on innovation in Europe. Mention has been made of the scarcity of charcoal and water power in Britain because of the stimulus it gave to the use of coal and steam power. The sharp decline in the whale catch after 1850 provided the United States with an impetus for discovering and refining petroleum as a domestic fuel. It was the French and the Germans, respectively, who benefited most from scarcity of natural soda ash and nitrate fertilizers, leading to the development of synthetic soda ash and synthetic ammonia. The British at the time had plenty of secure sources of imported natural raw materials from their colonial empire. Germany also pioneered in the development of synthetic fuels, synthetic rubber, plastics, and synthetic detergents, for similar reasons.

The role of scarcity is particularly well illustrated by the case of aniline dyes. W.H. Perkin an English chemist, accidentally discovered how to make a synthetic mauve dye from coal tar while seeking a way of synthesizing quinine to protect British rubber planters in tropical colonies from the source of malaria. Perkin immediately saw the potential for synthetic dyestuffs—along with others. By 1862 there were about 30 synthetic dye firms in Western Europe, mostly in Germany. The Germans went on to develop an enormous and powerful chemical industry. Despite Perkins's fast start, the British lagged in exploiting the technology of coal tar chemistry, mainly because they had large indigo and other dyestuff plantations in India.

Throughout the first and second industrial revolutions, the United States was a country rich in material and energy resources. As a consequence our industrial past owes much to labor-saving innovation but very little to innovations induced by resource-scarcity—in sharp contrast to Germany, for instance. In the third industrial revolution, now beginning, the role of scarcity is likely to be changed, if not reversed. The United States is no longer self-sufficient in resources: We are the world's largest importers of petroleum and most metal ores. The technological innovations that are likely to be called forth by this shift are discussed in the next chapter.

Important as scarcity has been and still is as an inducement to innovate, it seems that new opportunities created by organized science are becoming progressively more important in human affairs.

The narrowing U.S. lead in high technology is owed not to scarcity of any kind but to a cluster of discoveries in applied physics and electronics that were rapidly exploited in the 1950s and 1960s—largely because of encouragement by the U.S. government. These technologies are beginning to permeate every aspect of life.

Where does the U.S. economy go from here? Important new technologies are on the way. Whether the United States reaches the year 2000 as the world's leading producer of ICBMs and soya beans and nothing-much else, remains to be seen.

NOTES TO CHAPTER 4

1. Sand casting, invented by Darby in 1707 is the process of casting molten iron in a mold of compressed sand, using a binder to retain its shape. This process is still widely used, as in making automobile engine blocks. The Darbys later built the first cast iron bridge.

   Hammer forging is the process used by blacksmiths of shaping and simultaneous removal of impurities by working hot metal. Modern methods could not be used because furnaces of the time were barely good enough to melt pig iron (4 percent carbon), which melts at $1,150°C$, but not good enough to melt pure iron—or steel—which melts at $1,500°C$. Thus the conversion of pig iron to either pure (wrought) iron, or steel, was extraordinarily difficult.

2. As opposed to atmospheric pressure steam, employed by Watt.

3. This is astonishing, in view of the fact that most early boilers were still made of wood.

4. Another example of dramatic increases in productivity is given by Adam Smith in *Wealth of Nations*. When pin-making was organized as a cottage industry each pin-maker could make about 20 pins per day. Reorganized in a simple factory, with help from some crude jigs and tools, output increased to 2,400 pins per man-day. The price of pins fell dramatically, and demand rose to match.

5. For instance Sir William Siemens moved to England permanently in 1844 because British patent laws were more favorable than German laws. On the other hand most inventors got little monetary reward, unless they were also entrepreneurs.

6. But not always and equally so: for example Spanish imperialism in Latin America was totally destructive, both for the native population and for Spain itself. On the other hand the British did some quite good things in India, including unifying several hundred small principalities, introducing an effective civil service, a system of secular justice under law, forms of

representative government, infrastructure, and the beginnings of modern industry.

7. A zero-sum game is one in which the gains of one player are necessarily at the expense of another, because the total of all possible winnings is fixed. "Monopoly" is an example of such a game. However economic systems are much more complex than the game of Monopoly. It is possible for everybody to gain—or lose.

8. Many of them also greatly enriched themselves through their control over internal Indian trade.

9. Mercantilism required that imports be paid for by exports in kind, while exports had to be sold, at least partly, for money (bullion). Mercantilism was widely supposed to be responsible for the accumulation of investable capital in Britain during the seventeenth and eighteenth centuries.

10. To provide logistical support for these territories Britain began building a railroad over extremely difficult terrain, from Mombasa on the Indian ocean to Lake Victoria. The railroad was finally completed in 1903, at a cost to the British Treasury of £5,317,000.

11. Other factors contributed to the arms race, including increasing technological sophistication resulting in longer lead times. (See Richardson 1960.)

12. To keep the account from getting too long, I will not discuss some other advances in science and technology that also took place during this period. Among the more important: major innovations in industrial chemistry, electrometallurgy, the bicycle, the airplane, mechanization of agriculture, hybrid seeds, anesthesia, Pasteur's medical discoveries, and so on.

13. Spiegeleisen (mirror iron) is an alloy of iron, carbon, and manganese, used in the Bessemer steel to add both carbon and manganese, in a controlled manner, to the molten decarburized iron. The original Bessemer process involved an oxide (acid) liner for the converter vessel. The liner material did not combine with undesirable phosphate impurities, leaving them in the pig iron. S. G. Thomas and P. C. Gilchrist solved this problem in 1875. Their innovation was to replace the oxide liner material by a calcium carbonate (basic) material, which reacts with the phosphates and removes them.

14. The smelting of iron and the decarburization of pig iron both involve the oxidation of carbon (originally from coke), which generates a great deal of excess heat. Some of this heat can be recovered by passing water through pipes in the furnace walls. The water is heated and turns to steam, which can be used in turn to heat the incoming air.

15. The term *atmospheric* means the fuel-air mixture was burned at atmospheric pressure. The fuel was illuminating gas (coke-oven gas) obtained as a by-product of manufacturing coke from coal.

16. Britain was equally poor in land and energy sources within its own boundaries, but its effective control of the ocean trade routes gave Britain

easy access to energy resources from its empire and North America. France had less coal than either Germany or the United Kingdom but much more hydroelectric power, much better farmland, and large North African and Indo-Chinese colonies from which to obtain resources.

17. Today's analogy may be that two Japanese firms, Yamazaki and Fujitsu–Fanuc, have recently unveiled factories that function with very few employees and are completely unmanned during the third shift.

18. Nonsectarian tax-supported education in the United States has been generally regarded as essential to the functioning of a democracy, but the scarcity of skilled labor may have been a contributing factor. It was strongly supported in New England as early as 1750. The movement spread to the rest of the northern states in 1825–1830. Most states changed their constitutions to eliminate sectarianism in the 1840. The public school system was extended to high schools and college after 1840. The University of Michigan opened in 1841 as a public institution. By 1860 there were 17 state-supported colleges of a total of 246 colleges in existence at that time.

19. Notwithstanding a number of memorable episodes of labor strife.

20. Ford's wages ($5/day) were high enough to annoy other employers. But even average wages in the United States were much higher than in Europe. Americans feared competition from low-wage German competition.

> An American firm, it is said, will devise a new machine, and an export of the machine itself, or of its products, will set in. Then some German will buy a specimen and reproduce the machine in his own country . . . not only will the exports cease, but the machine itself will be operated in Germany by low-paid labor; and the articles made by its aid will be sent back to the United States. Shoe machinery and knitting machinery have been cited. (Taussig, 1915, quoted by Cooper, 1971).

5 THE NEXT INDUSTRIAL REVOLUTION
Petroleum Substitutes and
New Materials

Just as the first two industrial revolutions involved the development
of new sources of energy and new materials, so will the third. The
heyday of coal—which peaked just before World War I—was cut
short not by the exhaustion of coal supplies, but by rapid growth in
discovery and consumption of even more convenient fossil fuels,
petroleum, and natural gas. Fossil fuels accounted for nearly 75 percent of U.S. energy consumption by 1920 and close to 95 percent by
1950, as Figure 5-1 shows. Petroleum and gas surpassed coal immediately after World War II and peaked at over 75 percent in 1971.
The dominance of oil and gas in the U.S. supply picture has by no
means ended. However, a driving force behind the coming changes
is the fact that the United States is no longer a self-sufficient exporter of petroleum and minerals. Instead of a comfortable balance-
of-payments surplus in raw materials and cheap energy for domestic
industries, the situation on both counts has changed dramatically
in the course of the last three decades.

U.S. domestic petroleum production peaked in 1969-1970.[1]
Since then it has declined slowly despite stepped-up exploration. In
1957, when the Suez crisis caused a temporary halt in oil shipments
from the Middle East to Western Europe, the United States was able
to increase Texas output enough to make up the deficit. But when
the Arabs embargoed oil shipments to the West in support of Egypt
in the 1973 Sinai War, there was no surplus capacity to fill the gap.

Figure 5-1.   Energy Mix: United States.

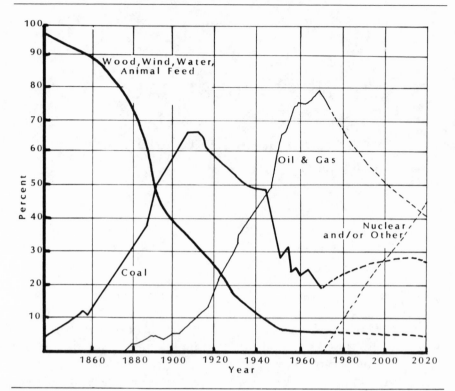

Source: Schmitt (1981).

When the embargo ended, the price of oil from the OPEC countries had been tripled—then quadrupled. By 1974 it was clear that the age of cheap energy in the United States was over. Since then there have been a number of false starts at the federal government level, including Nixon's abortive Project Independence and the Carter administration's Synfuels Corporation. Both programs were strongly resisted and finally defeated by a peculiar alliance of energy producers and consumerists. But major changes are going on elsewhere, both in the supply and demand areas.

CONSERVATION-INDUCED INNOVATION

It was once widely expected that nuclear electricity generated by breeder reactors would eventually be a cheap substitute for oil, with fusion power to follow in the early part of the next century. These hopes seem dim today for a variety of technical and environmental reasons. It is not that the technical problems are inherently insoluble, but the difficulties, environmental risks, and costs are immensely greater than had been expected in the 1950s and 1960s. Private investment in the nuclear industry has virtually ceased.

Synfuels, too, seem much more remote today than they did in the early 1970s. Exxon has withdrawn from its joint venture with TOSCO, shutting down the next-to-last active oil shale project in the United States. Most of the coal gasification and liquefaction plants that were being planned in 1979–80 have since been abandoned. Again the difficulties, risks, and costs were severely underestimated.

Part of the reason why the initial attempts at a quick technological fix of the energy problem—really a petroleum problem—have proved to be economically unattractive is simply that other price-induced responses have turned out to be cheaper, faster, and more effective. These have included stepped-up exploration and drilling for hydrocarbons, both in the United States and in other parts of the world. One consequence of this activity has been the unexpected discovery of quite a lot of natural gas in the United States. Also the sharp oil price increases in 1979–80 have led to the use of enhanced recovery techniques, which by themselves substantially increase proven reserves. Perhaps the most significant price-induced response has been *reduced consumption* of commercial hydrocarbons, due to

- Increased efficiency of energy use both by industry and consumers, as wasteful practices were eliminated, insulation was upgraded, and so on.
- A variety of decentralized energy sources, such as solar water heaters, efficient wood-burning stoves, windmills, and small "low-head" hydroelectric facilities.

The potential for energy conservation was grossly underestimated in the past.[2] Typical of the time, a report issued by Chase Manhattan Bank in 1972 said: "An analysis of the uses of energy reveals little scope for major reductions without harm to the nation's economy and its standard of living. . . . There are some minor uses of energy that could be regarded as strictly non-essential—but their elimination would not permit any significant savings." (Winger et al. 1972). This assessment has been sharply contradicted by subsequent history.

The combined effect of various price-induced energy conservation measures introduced since 1974 has reduced the commercial energy required per dollar of GNP in the United States by 10 percent from 1973 to 1978 inclusive, (Savitz 1981) and probably even faster since then—as compared to a slight (1.5 percent) *increase* in the previous decade. U.S. residential energy use from 1970 through 1980 increased by only 3 percent in absolute terms but declined (in effect) 18 percent as compared to what it would have been otherwise—without conservation (EPRI 1982). Because the United States consumes such a large fraction of the world's energy, world demand for petroleum has not increased nearly as fast as was expected a few years ago. Each year since the early 1970s, projections of energy consumption during the rest of this century have declined.

Figure 5–2 illustrates this quite dramatically for the case of electricity. As a result, despite the Iran-Iraq war, which cut petroleum exports from both countries, there was even a glut in 1982 and 1983. The glut will end eventually, but many irreversible investments in energy conservation are already in place and even more are on the way. The U.S. automobile industry's shift to fuel-efficient small cars is an example. New cars are, on the average, much more efficient than the vehicles they replace, which means the automobile fleet as a whole will continue to increase in overall efficiency for many years to come.[3] In 1969 the average U.S. car in service and the average new car obtained around 14 miles per gallon (mpg). By 1979 the average new car sold in the United States achieved 20 mpg. This figure is expected to rise rapidly in the 1980s to 33 mpg in 1985 and 39 mpg in 1990. It is expected by some industry experts that 50 mpg will be achieved by the average new car in the year 2,000 (Agnew 1980; Shackson and Leach 1980). Conservation alone cannot be a total long-term substitute for fossil fuels, however.

THE PROBLEM OF GASOLINE

Natural gas is now much more available than was expected in the 1970s. This is because a few years ago, when gas was thought to be growing scarce, distributors signed long-term contracts with producers calling for sharply increased future prices following scheduled deregulation. These higher prices are now being put in effect, despite the glut of gas, and their impact has been to induce residential and

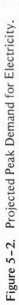

Figure 5-2.   Projected Peak Demand for Electricity.

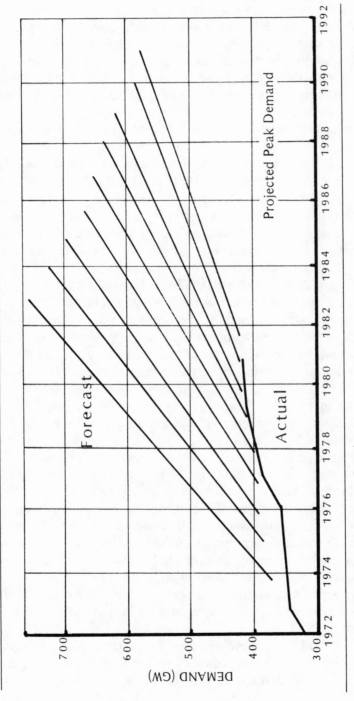

Source: Electric Power Research Institute (1981).

industrial consumers to invest in energy conservation via thicker insulation, more efficient boilers, double-glazed windows, and so on. Most of these improvements will be permanent. What all this means is that some industrial users of gas may be able to co-locate with natural gas fields that are "frozen out" of the big distribution systems and, thereby secure long-term supplies at favorable prices.

The most likely use for any excess natural gas is as a feedstock to manufacture methanol (methyl alcohol) as a substitute for gasoline.[4] Methanol is already manufactured in very large quantities from domestic natural gas by several well-established processes. It sells at around $1 per gallon. Imports from the Middle East or larger scale domestic production will probably force this price down during the coming decade. In terms of energy content, 2 gallons of methanol are equivalent to 1 gallon of gasoline. If the methanol were used as a substitute for gasoline in a conventional gasoline engine it would take 2 gallons of methanol to drive as far as 1 gallon of gasoline would go. However, methanol has a higher octane rating than gasoline and can therefore be burned at significantly higher compression ratios without knocking. Higher compression means higher thermal efficiency. General Motors (1982) estimates that a high-compression methanol engine would be 20 percent more efficient in thermodynamic terms than a present-day gasoline engine. Higher petroleum prices would hasten the substitution, but many industry analysts expect methanol to begin making inroads in the market around the year 2000, if not sooner, without government intervention.

Methanol is inherently a clean fuel. A conventional methanol-burning engine generates very low levels of emissions of carbon monoxide, hydrocarbons, and NO_x. On the other hand incomplete combustion produces aldehydes, which can cause eye irritation. Hence some sort of catalytic exhaust gas treatment would still be necessary, though perhaps a less elaborate one than the present system. A methanol-burning auto would also require a larger fuel tank, and some of the materials now used in the fuel system are subject to corrosion or otherwise adversely affected by methanol. Nevertheless the required alterations to the vehicle design are relatively minor. In fact, cars made by GM's German subsidiary (Opel) will be methanol-proof by 1985.

By the 1990s a more exciting possibility is likely to emerge: the methanol fuel cell. According to recent reports, an electric automobile based on such a fuel cell would be capable of over twice the thermodynamic efficiency of a gasoline engine. This means such a car could travel as far on a gallon of methanol as a present-day model could travel on a gallon of gasoline. While the fuel cell is still too expensive by a factor of 3 or so, its cost would certainly drop sharply with further development and large-scale production.

The most difficult aspect of transition to methanol as a fuel will be to ensure its availability in service stations as methanol-fueled cars are introduced. A similar situation has been faced and surmounted once before, when lead-free gasoline was required for new cars with catalytic mufflers, in the early 1970s, but only because of government intervention.

It is significant that, a few decades hence, as U.S. natural gas supplies eventually begin to phase out, methanol can be made quite easily from coal or lignite at only slightly higher cost than from natural gas. Two coal gasification projects now under construction expect to be able to product methanol at $0.35 to $0.37 per gallon, corresponding to a pump price of $0.60 or so. This may be optimistic. Nevertheless, it is pretty clear that long-term availability of liquid fuel from domestic sources now seems to be assured.

The economic implications of all this are actually fairly revolutionary. The small, lightweight methanol-fuelled electric car of the year 2000, capable of up to 60 mpg (gasoline equivalent), will be radically different from the large six- or eight-cylinder internal combustion engine powered vehicle typically produced in Detroit in 1970. The U.S. petroleum-refining industry will have to be dramatically restructured to accommodate the vanishing demand for gasoline. Crude oil will increasingly be used mainly as a source of petrochemical feedstocks, diesel fuel, and aircraft fuel. The need for petroleum imports should continue to decline and eventually cease—greatly extending the life of existing world petroleum reserves and reducing upward pressure on world prices.

For the U.S. steel industry, the implications are also quite dramatic. Domestic demand for structural steel (for automobile frames) and sheet steel (for auto bodies) will continue to drop as other structural materials such as plastics continue to increase in importance relative to steel. Much of the so-called integrated steel industry will have ceased operations in the United States by the year 2000, especially the older and smaller facilities without access to deep water transportaion.

OTHER MATERIAL SUBSTITUTIONS

The chief factor inhibiting the substitution of either plastics or aluminum for steel in automobiles is the price, which generally remains higher than the price of steel, because the alternative materials embody so much more expensive energy. It is unclear whether consumers realize that using light material instead of steel generally results in a significant fuel saving over the lifetime of the vehicle and are willing to pay more in capital (and finance) costs to save money over the life of the car. While many buyers are not so receptive to such trade-offs, they might become more so if interest costs declined or fuel costs or fuel taxes rose.

In regard to ways of bringing the cost of aluminum down, the history of industrial chemistry is instructive. It has been an evolutionary process whereby waste products are eigher eliminated by process substitution or uses for them are found (Multhauf 1967). Chlorine, for instance, was once a low-value by-product of the manufacture of sodium hydroxide. Its availability stimulated research, which led to the development of a variety of useful chlorinated hydrocarbon solvents and plastic and elastomers. More recently blast furnace slag has found so many uses that old slag heaps are even being "mined" in some locations. The next waste product that will probably find significant new uses is fly ash, from the combustion of bituminous coals in electric power generation plants.

Fly ash is a variable mixture of metal oxides, consisting mainly of silica alumina (20–25 percent) and iron oxides (15–20 percent). Ash from coal-burning boilers has further useful attributes as industrial feedstock; it is finely divided, of relatively uniform composition and size distribution (in a given location), and it is free of water.[5] Ash is also available in very large quantities—up to 450,000 tons/year from a single large coal-burning electric generating plant, and around 100,000,000 tons/year for the nation as a whole.

It is easy to calculate that if the fly ash now generated in the United States could be reduced to its constituent metal oxides, it would easily replace 100 percent of the imported aluminum ore (currently bauxite) and 40 percent of the imported iron ore. The question arises, Will it be technologically and economically feasible to utilize fly ash as an ore during the next 20 or 30 years?

Actually a number of physicochemical processes to extract alumina from ash have been investigated by various research groups, including the Bureau of Mines and the Electric Power Research Institute (EPRI 1982) in the United States as well as by groups in Poland, South Korea, Japan, and elsewhere. Some of these experimental processes appear in theory to be economically competitive with the conventional Bayer process for extracting alumina from bauxite, provided the alumina extraction process can be integrated into a larger complex of interrelated processes, usually including at least an electric power plant generation and a Portland cement plant (to utilize the excess silica). The difficulty in this, of course, is that the capital requirements for a viable complex would necessarily be very large. On the other hand such an integrated complex would, ipso facto, generate much less net waste and pollution than would be produced by independent facilities not linked together.

There are further technological possibilities on the horizon that could bring the costs down further. Recent developments in electrochemistry suggest, for instance, the possibility of a sequence of linked bipolar electrolytic cells using inert electrodes. Fly ash dissolved in molten cryolite would enter at one end, while pure metallic iron, silicon, and aluminum could be continuously deposited on and removed from successive electrodes at controlled voltages.[6]

It is worth pointing out, incidentally, that lunar soil is rather similar to fly ash in composition, except for a slightly higher titanium content. The electrolytic extraction scheme would be extremely appropriate for a future materials extraction plant located on the lunar surface (Jarrett et al. 1980). This is so because, in space, energy—either electric or thermal—is readily available, but water-based or halogen-based chemical processes of any kind would be very difficult to implement because of the recycling requirements. Such a lunar extraction plant could be the basis of a future space-based silicon solar-cell manufacturing industry, for instance.

A lunar materials plant would produce aluminum, titanium, and ultrapure silicon. Thus another long-term possibility is widespread direct conversion of solar energy to electricity via photovoltaic cells deployed both on earth and in space. Decentralized roof-based photovoltaic units tied into the existing electric power distribution system would be very attractive to many owners of multifamily residences and office buildings, especially for purposes of air conditioning and operating office equipment. The advantage of placing large solar energy collectors in space rather than on earth is that they can operate 24 hours a day, without atmospheric interference. This yields a sixfold increase in electricity output per cell as compared to a similar collector on the ground. Very little of this advantage would be lost if the energy then were converted to microwaves and beamed to earth. However, the disadvantages of operating in space are still quite formidable. A solar power satellite would necessarily be large and expensive to build, even if most of the construction materials were themselves produced on the lunar surface from lunar materials. It would also be vulnerable to sabotage or attack. The microwave beams, covering hundreds of square miles, might be hazardous to birds and other living organisms on earth, and the network of receiving antennae would take up a lot of space on the ground.

SOLAR ELECTRICITY

The economic feasibility of electric power from the sun is not yet established. It depends on reducing the costs of photoelectric cells and systems by a factor of 10 or so from 1980 levels.[7] This seems like a lot to expect at first sight, but it would not be astonishing at all in the light of past experience with other electronic products such as semiconductors. Indeed, a very sharp reduction in the cost of ultrapure silicon crystals as the scale of output increases is exactly what one would anticipate from the *experience curve* mentioned in Chapter 3 (Figure 3-5).[8] Alternatively, cheaper polycrystalline or even amorphous forms of silicon are feasible, though slightly less efficient at converting sunlight to electricity. In any case the technology is rapidly progressing and success appears to be virtually certain in the not distant future.

Worldwide sales of photovoltaic cells and associated equipment were only $6.8 million in 1979, rising to $39.3 million in 1980, $57 million in 1981, and $83 million in 1982. The market should reach the $1 billion level in the late 1980s. Much of the present civilian market for high-cost solar electricity is in foreign countries, especially in remote places where electricity must otherwise be generated by diesel-powered generators. The U.S. market is mostly for highly specialized purposes such as satellites. U.S. firms still supply 60-65 percent of the world market, but major Japanese penetration seems virtually certain as soon as true mass production of solar cells begins.

It seems likely that there will be a major growth industry based on manufacturing solar cells and related equipment before the beginning of the next century. The major question is whether that industry will be primarily based in the United States or in Japan.[9] One factor in favor of the United States is the strong link between photovoltaic technology and space industrialization, where the United States is still the technological leader. But, as in other fields, Japan appears to be moving ahead faster.

NEW MATERIALS

A number of so-called exotic materials that have been developed primarily for aerospace applications seem to be on the verge of significant commercial use. One of them is titanium, a metal that has tensile strength comparable to steel and much greater than aluminum but is lighter and more corrosion resistant than steel. Titanium is not rare—it is the fourth most abundant of all metals—but it is very costly to refine from ore and to form into useful products. Titanium obviously has major uses in high-performance aircraft, where it has no real substitutes, but that market has been cyclic and unreliable. Other potential markets primarily as a substitute for stainless steel in chemical plants, water desalination plants, and power stations have been slow to develop. Although the titanium industry first developed in the United States, the most active development is now taking place in Japan, which now produces half of the titanium used in the non-Communist world—much of it exported to the United States.[10] However, titanium now costs more than ten times as much as aluminum per pound, and is not likely to compete with aluminum as a structural material as long as it is produced on the earth's surface. This situation could change, however, if a lunar-based electrolytic-materials-processing industry were to be created in the next century.

Another category of exotic materials is actually one of the oldest known to man: ceramics. What is new is the potential use of ultra fine-grained ceramics for high-performance jet engines or for automobile engines. Apart from utilizing low-cost materials, such engines would be able to withstand operating temperatures several hundred degrees higher than current metal engines can take. Higher operating temperatures mean significantly increased efficiency—that is, more miles per gallon. This possibility has been under study in U.S. research laboratories for many years and the U.S. Defense Department is supporting research on a ceramic gas turbine engine scheduled for the 1990s, but again Japan seems to have taken the lead. Kyoto Ceramics Company (Kyocera), the world's largest industrial ceramic manufacturer, recently displayed a prototype ceramic-metal automobile engine. While mass production is still several years away, the prospects are favorable. One Japanese stockbroker has estimated that Japanese demand for ceramic engines would reach $1.5 billion for automobiles and $1.35 billion for other vehicles by 1990 (Gregory 1982).

Plastics are not as exciting as they were when the protagonist of "The Graduate" in the mid-1960s was solemnly advised to seek his fortune thence. But one rapid growth area is that of glass and carbon fiber-reinforced plastics—used in a variety of aircraft parts and sporting goods (golf clubs, tennis rackets, fishing rods, baseball bats, and football helmets). Future uses will be in automobile exterior body parts, wheels, springs, brake linings, and even gears. These materials are much more expensive per pound than steel or aluminum, but they are much easier to form into final shapes. Thus fiberglass auto bodies, for instance, are only slightly more expensive than steel bodies and much lighter. In fact, the newer carbon-fiber-reinforced plastics have considerable potential as synthetic metals because of their high strength, light weight, and resistance to corrosion. The U.S. market for this material has been growing at a spectacular pace, from 160 tons in 1978 to an estimated 21,000 tons by 1985 (Gregory 1982). Japan produces an estimated 65–80 percent share of the world output—most of which is currently exported to the United States.

MATERIALS FROM SPACE

It will be recalled that 1969 was significant in part as the year Americans first walked on the moon. The original Apollo project was perhaps intended mainly to enhance national prestige, but it did give the United States a useful technological capability. The successful Explorer and Voyager projects of the 1970s and the first flights of the Columbia space shuttle starting in 1981 have kept the U.S. moving forward in this field. But the USSR has also moved forward rapidly, and both Japan and France are also becoming strong competitors in space technology. Without question the first large payoff from space technology has been in satellite communication. Revenues of around $1 billion from this application alone were realized in

1977. This should rise rapidly to more than $10 billion by the year 2000 and perhaps $80 billion by 2010, according to NASA. Information services will unquestionably be the primary justification for investment in space-related endeavors for the next 20 years. However, energy and materials from the moon or asteroids may very well become the major export product of space industrialization before the third industrial revolution runs out of momentum sometime in the first quarter of the next century.

The reason for investing in R&D oriented toward space manufacturing is, basically, that significant exploration and utilization of the solar system and the space environment—for whatever purpose—depends upon the existence of a substantial base of operations on the moon or in high earth orbit. Such a base in turn would require substantial life-support and manufacturing capabilities. It is clear for reasons of energy cost that the vast bulk of the necessary materials and energy *must* be obtained from the moon or the asteroids. Lifting materials from the earth's surface is simply not realistic. It follows that highly automated materials extraction and fabrication capabilities are also a prerequisite to the industrialization of space.

NOTES TO CHAPTER 5

1. This is one reason for using 1970 as the starting date for the third industrial revolution. Another milestone was the invention of the microprocessor.

2. Part of the reason for the tendency to underestimate the potential for conservation is the pervasive misconception of the efficiency of energy use in practice. Widely publicized studies such as *Understanding the National Energy Dilemma*, prepared and circulated by the Congressional Joint Committee on Atomic Energy (1973), asserted that "lost energy" amounted to slightly below 50 percent of the total energy consumed in the United States—implying an actual efficiency of more than 50 percent of theoretical maximum. Examination of the assumptions on which this figure was based reveals, however, that many sources of loss were ignored, and the term "useful energy" as employed in the JCAE Report is virtually meaningless. In fact the actual efficiency of energy use in the U.S. economy was, and probably still is, closer to 5 percent than 50 percent, even allowing for recent improvements (Ayres, 1977). There is still a great deal of room for further price-induced conservation in the future.

3. So far, significant savings in fuel consumption have been made by paring the size and weight of the car, cutting tire and wind resistance, and introducing computer-controlled carburetion. Future improvements from more

efficient engines, "infinitely variable" transmissions, and use of light-weight materials (plastic, aluminum) have yet to be introduced.

4. Compressed methane can also be used directly, in gaseous form, but this is probably only feasible for bus, truck, or taxi fleets. The chief motivation for using methane directly is that it costs much less than gasoline per unit of energy stored.

5. Unless, of course, it is captured by means of a wet scrubber. However, if there were an attractive economic use of fly ash, dry electrostatic precipitators could be utilized exclusively.

6. Obviously the continuous electrolytic extraction process envisioned here is quite speculative. It builds, however, on the new bipolar electrolytic cells, with inert electrodes, that are being developed by Alcoa Research Center. I am indebted to Noel Jarrett for his ideas on this subject.

7. Present commercial photovoltaic solar collector systems cost about $15 per watt of continuous power output (e.g. Arco-Solar). DoE's formal research goal is to achieve system costs of $0.70 per watt, which would make photovoltaic systems competitive with oil-burning central power-generating plants. However, even at $1.50/watt the potential market is likely to be extremely large. Japanese manufacturers anticipate achieving a cost of 1,000 yen per watt by 1984–85 (roughly $4 to $5, depending on the dollar-yen conversion rate).

8. The experience curve or learning curve is, of course, an empirical phenomenon based on past experience with successful products, not unsuccessful ones. There is no a priori guarantee that photovoltaic cells can actually be produced economically on a large scale. At present some serious technical difficulties remain. The fact that silicon chips can be mass-produced nowadays at very low prices does not itself guarantee that silicon photovoltaic cells can also be produced cheaply. The reason is that chips are essentially two-dimensional objects with circuit elements implanted on the surface by a process akin to photoengraving. Photovoltaic cells, however, are presently made from three-dimensional ultrapure silicon crystals converted into semiconductors by a process of diffusing desired impurities through the outer surfaces. This diffusion process requires heat as now performed, and it is hard to control and unreliable. The loss rate therefore is high and hence so is the unit cost. The kind of cost reduction that seems to be required would not arise from increased production scale or learning alone—a major breakthrough is needed. Fortunately there is reason to believe silicon crystals can be grown with the requisite impurities already in the structure, thus bypassing the tricky diffusion process. (Neary et al. 1982).

9. Other countries are lagging but cannot be ruled out.

10. Notwithstanding the fact that titanium is even more energy-intensive than aluminum.

6 THE NEXT INDUSTRIAL REVOLUTION
Electronics, Information, and Flexible Automation

The observation that an *information*-based technological revolution is under way is not new. Sociologist Daniel Bell, among others, saw clearly by the mid-1960s that the United States was becoming a postindustrial society based on the production services, especially information-related services (Bell 1967, 1973). The third industrial revolution is based in large part on advances in telecommunications and information processing. Manufacturing and construction employment has remained more or less static in absolute terms for many years, but the work force as a whole grew tremendously in the 1960s and 1970s. Long-term employment trends, as illustrated in Figure 6-1, show clearly that the major growth sector for the past century has been in information-related areas. At first this category comprised mainly the one-way communications media—books, magazines, and newspapers. The print media expanded in scope as a consequence of major improvements in high-speed printing technology during the last quarter of the nineteenth century. Telecommunication, beginning with the telegraph and followed by the telephone, became an important employer after 1900. Later economic growth came from new one-way media: radio in the 1920s and 1930s—when the advertising industry also expanded—and television in the 1950s and 1960s.

A series of technical developments in two-way (point-to-point) communications since 1850 have combined to increase channel

Figure 6-1. The Four Sections of the U.S. Labor Force, by Percentage, 1860–1980 (*Using Median Estimates of Information Workers*).

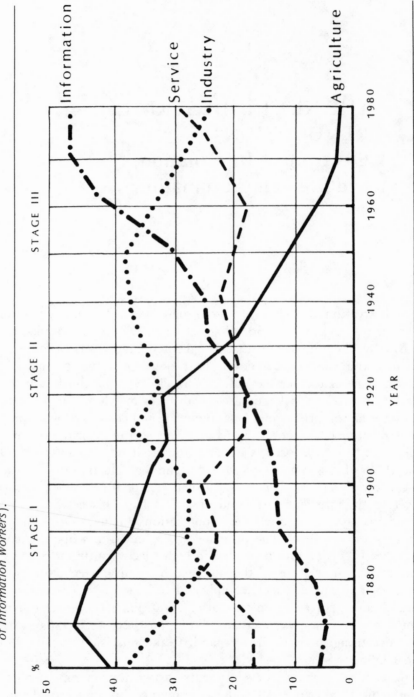

Source: Porat (1972).

capacity—bandwidth—by a factor of more than 100 million since the first telegraph experiments with an oscillating needle (Figure 6-2). The first telephone lines, constructed in the 1890s, required one wire for each distinct channel. Cities rapidly became festooned with overhead wires in the first years of this century. Multiplexing provided some relief: by 1916 Bell Laboratories engineers were able to squeeze 12 voice channels into a single carrier wire pair. Carriers became much smaller as amplifiers (repeaters) were improved. By 1946 coaxial cables could transmit 600 voice channels. The next step was the microwave waveguide—a by-product of World War II radar research—which increased channel capacity by a further factor of 3. The Bell system began constructing a system of microwave links between major cities in the United States in the early 1950s.

Since then technical progress in two-way communication has accelerated. The capacity of coaxial and microwave links built in the late 1960s reached 32,000 voice channels. Satellites soon became a part of the network, beginning with Bell Laboratories' Telstar satellite in 1962. But possibly the most important technological development of all was the critical invention in 1962 of the laser—a device to create and amplify coherent monochromatic light beams suitable for information-carrying purposes. This invention permits the substitution of bundles of glass fibers carrying signals modulated on coherent beams of visible light for copper wires and microwave guides. New York and Washington, D.C. are already linked by an experimental fiber-optical communication system. The first intracity fiber optics system is now being constructed in Pittsburgh by Pennsylvania Bell. An optical link connecting Boston, New York, and Washington capable of transmitting 80,000 telephone calls simultaneously is being planned by AT&T. The large-scale replacement of copper wires by fiber optic cables is accelerating. Optical technology also appears to be the key to the next generation of faster computers. Costs of sending messages have correspondingly declined by many orders of magnitude in the last hundred years. Dramatic cost reductions have resulted in enormous increases in two-way communications traffic.

One of the more significant trends of the 1970s has been the convergence of one-way and two-way communications media, beginning with cable television. The next step appears to be electronic newspapers. Cable systems are now beginning to acquire interactive capabilities, which will permit the user to interrogate the system or feed

Figure 6-2. Key Events in the History of Telecommunication.

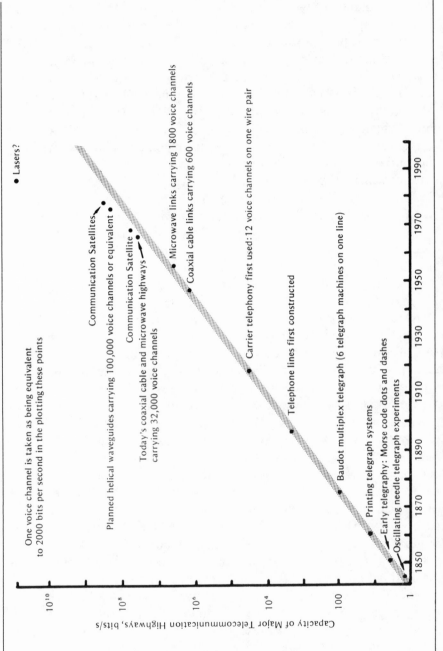

Source: J. Martin, Future Developments in Telecommunications, (Prentice Hall, 1971).

information to it. This capability opens the door to a variety of new services that are now being tested in several countries.

Of course the other important new information technology is the electronic computer. The first attempts to build a programmable mechanical calculating machine are attributed to Charles Babbage, an English mathematician for his "difference engine" (1823) and the unfinished "analytical engine" (1833–1971). While Babbage did not succeed in developing a viable mechanical calculating machine, his work was influential and he was in some sense the inventor of the modern computer—including the stored program, which was kept on punched cards. In fact the punched card (borrowed, in turn, from Jacquard's automatic loom of 1804) retained its key role as a data input and storage device until the 1960s. Babbage failed because he lacked the necessary precision components and because there was little or no demand for high-speed mechanically assisted computation at the time. Demand developed slowly for the next century, as did the supporting technology. The general theory of computers was developed in the 1930s by Coffignal (France), Zuse (Germany), and Turing (United Kingdom). The first hardware appeared in the early 1940s.

Konrad Zuse actually built the first electronic computer in Germany in 1941, but Zuse's research was not supported by the German government. After World War II Zuse continued to develop computers for scientific and commercial customers, but lacked the financial strength of the U.S. firms. His firm, Zuse AG was finally taken over by Brown Boveri in 1964. At Harvard University, Howard Aiken, under International Business Machines (IBM) sponsorship, built a series of automatic sequence-controlled calculators, the MARK I, II, III, and IV. These were electromechanical machines. The ENIAC, the first truly electronic computer, was built at the University of Pennsylvania during the period 1943–1946 by J. Presper Eckert and J.W. Mauchly, under contract to the U.S. army. It contained 18,000 vacuum tubes, and could perform seven instructions per second, a great step forward over Aiken's electromechanical machine, which required 25 seconds per instruction, and required as much as a day to set up a new program. ENIAC still had no high speed memory or stored program capacity. The EDSAC at Cambridge University and MADM at Manchester University in the United Kingdom were the first computers with a stored internal program (1949), followed by EDVAC, designed by Burkes, Goldstine, and von Neumann at the University of Pennsylvania. The MIT-WHIRLWIND, finished in 1952, was the first to exploit ferrite cores in place of cathode ray tubes (CRTs) for high-speed memory.

Meanwhile, Eckert and Mauchly founded a company (UNIVAC) to exploit the new technology and accepted a contract to build several computers for the U.S. government. The fledgling firm ran into severe difficulties and was taken over by Remington Rand Corp. in 1950. The first copy of UNIVAC I went to the Census Bureau (18 months late) to process the 1950 Census. Five others went to other government agencies. The first commercial delivery was in 1954. This success stimulated IBM, which unveiled its first commercial stored-program computer, the 701, in 1953. Not only was UNIVAC's early market provided by the U.S. government, so was IBM's. Throughout the 1950s government contracts underwrote 60 percent of the R&D expenditures at IBM and provided a comparable fraction of the sales.

The most significant invention in the history of computers was the transistor developed at Bell Telephone Laboratories by John Bardeen, Walter Brattain, and William Shockley (1947). This was the outcome of a deliberate attempt to find a cheaper, more efficient substitute for the vacuum tubes that were then part of every electronic circuit. The telephone system was then the largest consumer of electronic circuitry, especially for amplifiers and relays. Transistors were almost immediately utilized in computers (1954), however, and constituted the first of a series of technological breakthroughs that have since increased computer speed and performance by a factor of 10 every three to four years while cutting data processing costs almost as rapidly. From the earliest transistors, which were essentially solid-state equivalents of a single diode or triode vacuum tube, scientists have concentrated on miniaturization of circuit elements and multiplexing of functions.

The next significant development, led by Fairchild and Texas Instruments around 1960, was the shift from individually manufactured germanium and gallium arsenide semiconductors to thin-film metal-oxide-on-silicon (MOS). This facilitated miniaturization and large-scale integration (LSI) of complex circuits onto a single silicon chip. This, in turn, led to the development, by Intel Corporation, of the dynamic random-access-memory (D–RAM) in 1970 and the microprocessor in 1971. Both compressed more than 1,000 transistors onto a fingernail-sized silicon chip. Large-scale integration is now evolving into very-large-scale integration, or VLSI. Commercial

VLSI chips in 1983 contain as many as 64,000 elements, and chips with 256,000 transistors are already being produced both in the United States and Japan. By 1990 over 1 million transistors per chip will be standard. The Japanese are openly aiming to achieve 10 million. Costs have dropped as rapidly as microminiaturization has progressed (Figure 6-3).

Figure 6-3. Declining Cost per Bit, Computer Memories.

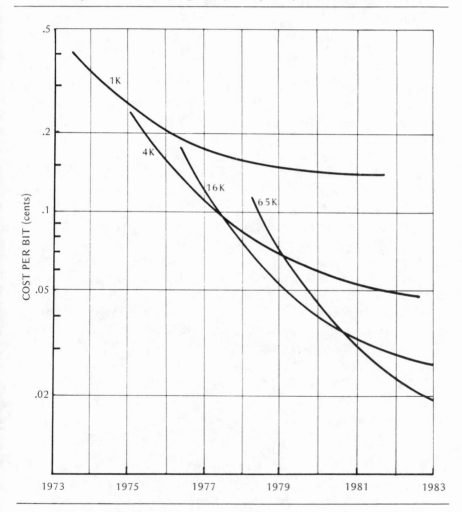

Source: Noyce (1977).

From 1951 until 1970 or so, computers appeared in four fairly well-defined generations, each an order of magnitude more potent in number crunching ability than its predecessor (Figure 6-4). After the first generation led by the UNIVAC I, the transistorized second-generation IBM 704 appeared in 1955. The Control Data Corporation CDC-3600 and its successor the CDC-6600 introduced the third generation of large computers in 1963 and 1964; the Burroughs 6700 led the fourth generation in 1969. In hardware terms the leading edge main-frame computers since then have been the Hewlett-Packard HP 21MXH (1974) and the IBM 3033 (1978). But since the mid-1970s it has seemed to be less appropriate to measure the state of computer technology simply in terms of speed, cost, and memory capacity, (all of which, especially cost, are directly related to miniaturization). Other subtler features, such as parallel processing, have become more critical. The Japanese government has established an ambitious project aimed at producing a fifth-generation computer by 1990.

In some respects the advent of the microprocessor has drastically changed the competitive environment for computers. The fastest growing segment of the industry in the 1970s was the minicomputers. Today the major growth is in microcomputers, small business computers, and personal computers. One obvious next step is for many of the mini's and micro's to be linked by telephone lines to large central processors and data banks, thus becoming "smart" terminals! The ultimate capabilities of such a marriage of computers and telecommunications are nearly unimaginable.

This convergence of information processing and information transmission technologies has already brought computer giant IBM into direct competition with the telephone giant, AT&T. The pending deregulation and breakup of AT&T is a direct response to this competition. The areas of overlap are indicated in Figure 6-5.

EARLY APPLICATIONS OF COMPUTERS

The development of a new technology alone does not create industrial (or social) revolution. The application of a new technology across a wide spectrum of societal activities is equally important. At the time of the initial introduction of commercial electronic computers in the 1950s, the applications of computers were largely of a

Figure 6-4.   Computer Performance Index, Selected Computers.

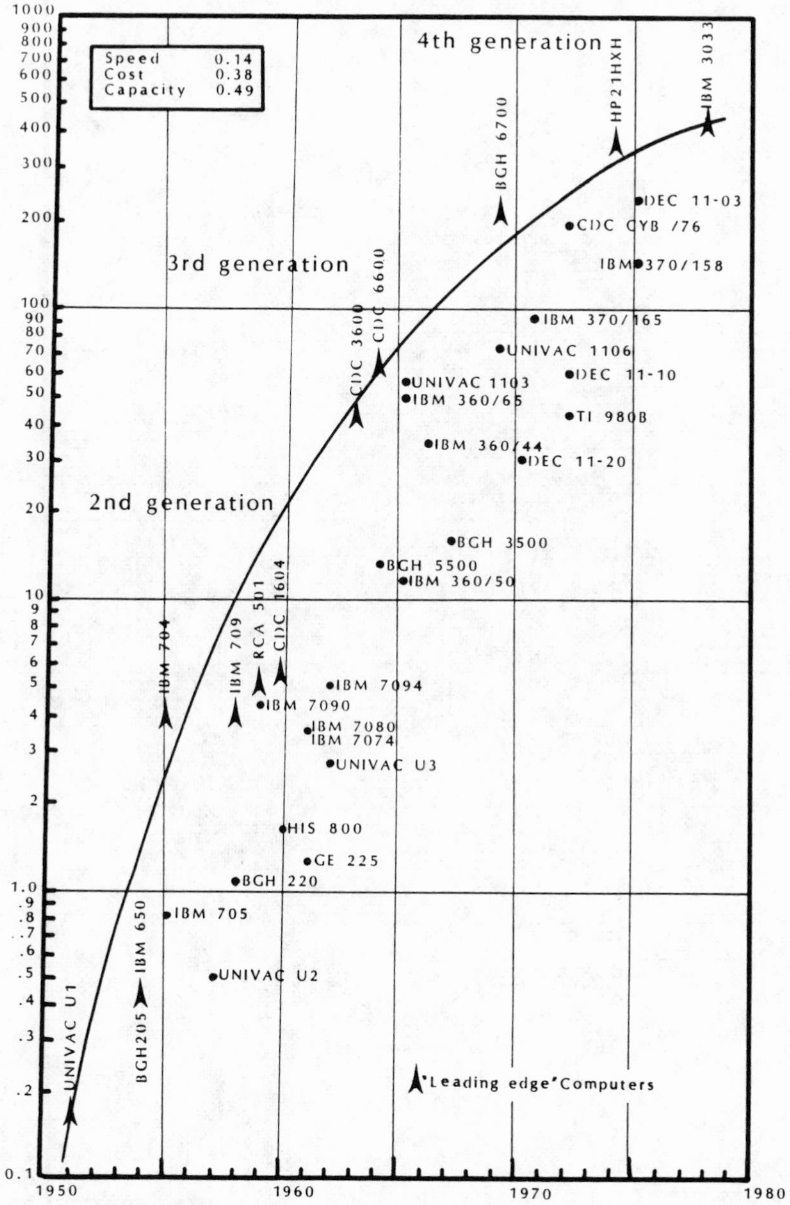

Speed	0.14
Cost	0.38
Capacity	0.49

Source: Gordan and Munson (1983).

Figure 6-5. Information-Related Businesses.

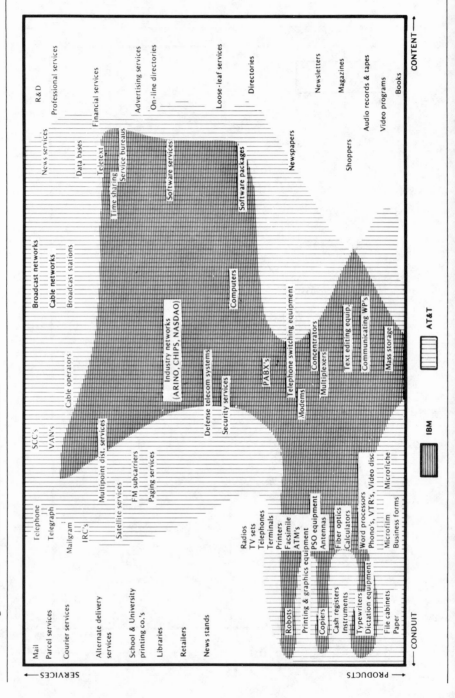

scientific, technical, or military nature. ENIAC, EDVAC, MIT–WHIRLWIND, and many later advanced prototypes were sponsored by the military. The first six copies of UNIVAC I were purchased by government agencies.

The ENIAC was originally designed to compute ballistic data for artillery. Early computers were extensively used in connection with the development of the H-bomb (especially the design of certain components), the design and operation of the early warning radar (AWACS) system, the targeting of intercontinental ballistic missiles (ICBMs), in design of high-performance aircraft, and in the deciphering of codes. The need for fast, massive computational capabilities has continued to grow and provides the primary driving force behind the development of the large top-of-the-line main-frame computers, many of which are purchased by nuclear weapons laboratories, NASA, National Security Agency (NSA), the Central Intelligence Agency, and the Defense Department and its major contractors. Weather and climate modeling and forecasting are another source of continuing demand for very large, fast computers.

The first UNIVAC I having been built for the U.S. Census Bureau, to process the results of the 1950 census, large-scale processing of statistical data from forms and administrative records soon became the second major area of application of electronic computers. Social security accounts, bank accounts, insurance policies of all kinds, department store charge accounts, brokerage accounts, hospital records, inventories, payrolls, pension records, tax records, legal records, automobile registration records, magazine subscriptions, advertising accounts, and a host of similar records are much more easily and quickly updated and accessed by computers than by file clerks. Searching, sorting, and compilation into categories are far easier and quicker with a computerized data file than with a manual file. Thus many tasks that were simply too expensive to do before computers came along are now routine. For instance,

- Magazine publishers, broadcasters and advertisers correlate sales in relation to advertising and promotional effort, by location, and by age, sex, income, and other known characteristics of readers or viewers.

- Automobile manufacturers now correlate new car registrations by model and location, with age, sex, income, and other known characteristics of buyers.

- Insurance companies are now able to compute actuarial tables and differential risks for a large number of categories of policyholders. This permits them to reduce losses by raising rates for high-risk groups and at the same time reduce costs of insurance to low-risk groups.

- Service firms and stores can better keep track of credit, identifying customers who consistently pay late.

- Firms on all kinds of business can keep detailed records of the habits, preferences, and other characteristics of their customers—past, present, and potential.

INFORMATION PROCESSING
AND NEW SERVICES

Once large computerized files of information existed, many new services came to be based on them. The buying and selling of specialized lists for marketing purposes is one simple example. Computer searches can be carried out systematically across a number of lists—as is done in credit checking. The modern credit-card industry could not exist without such computerized files, because of the large amount of record processing involved in each transaction compared to old-fashioned cash sales. A more complex service based on specialized list processing is computerized payroll preparation, even for very small businesses. The payroll clerk is now virtually obsolete. (This is one of the few categories of labor that were actually displaced by computers.) Computerized subscription fulfillment services for small magazines are still another example.

Centrally stored computer files began, in the 1960s, to be linked to networks of specialized input/output devices. The enormous computerized international airline reservation system, which required well over 10 years to develop from its primitive 1950s form (American Airlines–Teleregister Corporation) to operational status and another decade to reach its present level of sophistication, is one example. The modern airline industry could not operate as an efficient mass-transportation system without such a system. The benefits accrue to airlines in terms of higher equipment utilization and capital productivity, rather than reduced labor inputs.

Automatic bank tellers constitute a more decentralized application of this sort. The use of clerks for simple transaction processing (deposits and withdrawals) is sharply reduced and the automatic teller is open 24 hours a day. Point-of-sale terminals—optical scanners linked to electronic cash registers—are already linked to stores' central billing systems to authorize credit purchases and will soon be linked to banks for automatic debiting of customers' accounts. A completely automatic self-service gasoline station, utilizing special debit cards, should soon be open for business in California. While consumers may be slow to accept automatic debiting, this seems to be a wave of the future.

The linking of household television sets by a network of cables opens up a far more revolutionary possibility. It is already certain that the TV set, now strictly an output device, will shortly acquire input capabilities to permit viewers to interrogate the network. One of the first uses of this capability, called *teletext*, will be to allow users to request items from a specified menu of available information, like the Yellow Pages, price lists, entertainment listings, library card catalogs, financial data, and so on. Such a system has been operational for several years in Britain. The first U.S. teletext service began in the fall of 1982. Electronic newspapers, delivered to the TV terminal, are also in the planning stage and will be available in some areas within a year or two.

A truly interactive two-way capability, permitting the user to interact with the system or to feed information into it, by phone or keyboard, will follow within a few years. Such systems have already been operated on an experimental basis in a number of communities in the United States and other countries. Specialized computer-conferencing systems permitting a group of people in many locations to meet via terminals, and exchange information and ideas are already installed in some large multiplant firms, including Westinghouse. Eventual applications of two-way interactive systems will certainly include shopping (by debit or credit), public opinion polling, local and town meetings, census surveys and other functions of government, and possibly some elements of public education. Eventually citizens may also vote from their home terminals.

More elaborate interactive terminals capable of carrying wide-band video facsimile data in both directions will permit substantial numbers of individuals to work partly or wholly at home. People like stockbrokers, real estate agents, insurance agents, travel agents, designers, engineers, architects, programmers, secretaries, editors and indexers will be able to take advantage of this possibility. One of the benefits to society would be reduced need for peak hour travel to and from downtown areas, theoretically resulting in less congestion, less air pollution, and less waste of energy.[1]

THE OFFICE OF THE FUTURE

Offices will still be needed, of course, for a long time to come. In the first place much business depends on personal relationships that require face-to-face communication. It is an established and not very surprising fact that a video picture, no matter how good in quality, is not a viable substitute in all circumstances for direct contact.[2] Moreover, for many kinds of information processing there are still significant economies of scale favoring an office over a decentralized collection of remote work stations. The efficient use of expensive equipment (other than central processors), such as data storage (tape and disk), graphics terminals, composing, copying, typesetting, printing, and binding machines, optical scanners, microfiche/microfilm readers, provides further justification.

The office of the future is rapidly taking shape in response to rising labor costs and an explosion of demand for information and documentation. Office typewriters are still manually operated, but by the late 1970s many of them began to acquire internal memory units and simple programs to permit storage, manual editing, and automatic retrieval of typed material. The personalized computer letter, a straightforward application of the typewriter with a memory, has been a standard feature of direct mail marketers' and the offices of elected officials for well over a decade.

The next step, already nearly a decade old, was the stand-alone word processor. Basically, a microprocessor and a CRT screen were added to the memory typewriter. The material is entered via the keyboard and appears on the screen. Editing procedures are vastly simplified and automated, with the help of the microprocessor. For instance, one can instruct the machine to delete certain words or phrases whenever they appear, add move or delete sentences or paragraphs, and automatically renumber tables, figures, or references after such manipulations. Some sophisticated word-processing software systems even correct obvious misspellings.

The next step was to combine word processing and computing capabilities, to add a truly programmable computer capability, plus high-speed data inputs (such as a disk drive or telephone modem) to the stand-alone word processor. Such systems finally became technically and economically feasible with the advent of the personal computer, around 1980. And almost immediately, offices are beginning to link such units together into an integrated network, with direct access to central computerized files of customer records, employee records, service and maintenance records, sales, production, inventories, and so on. Thus central file data can be retrieved, processed as required (tabulated or graphed, for example) and incorporated into a report or memo being prepared by the operator.

It hardly needs to be pointed out that many existing office jobs will eventually be eliminated by these innovations. One of the first targets of the automated office is the central typing pool. Another is the filing clerks, with the user having direct access to computerized files, filing and data retrieval will be simplified and less subject to transcription error. Secretaries, too, will have less typing to do, as their bosses learn to do their own editing (if not their own typing). Many middle managers themselves will become redundant as the new technology reduces the need for humans to function (in effect) as information traffic controllers and interpreters for higher levels of management. The number of levels can be cut back.

Already there are reports of substantial reductions in staff directly attributable to office automation. One corporation cut its office staff by 45 percent (120 people), reduced office space requirements by 20 percent, and saved its principal officers 30–50 percent of their time, formerly needed for office housekeeping functions (Weil 1982). Considering that the technology is still in its infancy and that 50 million Americans are currently employed in office jobs, the potential social impact of office automation is likely to be far more pervasive than the potential impact of factory automation.

SCHOOL OF THE FUTURE

While some routine educational activities, such as drill and practice may be accomplished at remote locations via interactive cable networks, the major functions of schools, at all levels, will almost certainly continue to require some centralization. The fundamental reason is the same as that for the survival of some offices, namely that effective teaching and socialization of children requires direct contact in a common environment.

Nevertheless it appears certain that computers will play a growing and eventually pervasive role in the educational process. Many administrative functions in public schools can and should be computerized, now that computer power has become so inexpensive. Large amounts of data (attendance records, measures of student performance) are accumulated daily, weekly, and monthly by each teacher and much of this must be reported to the office of the principal. At this level there is some routine processing—compilation, sorting, and aggregation—for further transmission to the school superintendent's office. Again, after routine processing, information moves on to the local board of education and to the state office of education. Most of this massive administrative information flow could be automated now, and eventually it will be, based on inputs taken directly from terminals in each classroom.

The second important use of computers will be to facilitate skill development and routine instruction. An example of the first would be computerized programs, linked to specialized terminals, to teach manual skills involving eye-hand coordination. Typing, driving a car, and music reading or playing of keyboard instruments involve more or less extensive practice. The specialized terminal designates an exercise, monitors the practice session, records and (optionally) displays errors, and automatically grades the result. If the student performs above a suitable standard, a more difficult exercise is presented. The availability of such practice terminals should sharply reduce the cost of piano instruction, for instance, since the piano teacher could concentrate on fingering, technique, and interpretation, and understanding of the music itself.

Computer-aided-instruction (CAI) as applied to relatively routine subject matter is exemplified by the use of terminals to drill and test students in subjects like spelling, vocabulary, grammar, and reading comprehension in any language as well as arithmetic, algebra, geometry, symbolic logic, and so on. Such interactive instructional programs are more or less state of the art today. Their basic weakness is inability to diagnose patterns of incomprehension on the part of students and to respond appropriately. This would require more advanced capabilities, generally lumped under the heading *artificial intelligence*. Intelligent teaching programs are a major goal of current research.

This leads to the third use of computers in schools, which is to teach children how to use (program) and communicate with computers themselves. The usual term for this is *computer literacy*. The general idea is to give each child access to computers, from the first grade on.

The barriers are lack of investment in computers and software by school systems, the conservatism of professional educators—who are themselves largely unfamiliar with computers—and the relatively primitive state of educational software.

These barriers are mutually reinforcing. They will be broken down in time, however. Already Carnegie–Mellon University in Pittsburgh has contracted with IBM for a much more radical program, amounting to total immersion of students and faculty in an environment providing a variety of computer capabilities and services. By 1986 or 1987 it is planned that each of 5,000 students and 500 faculty members of Carnegie–Mellon University will have an advanced high-capacity personal computer that will also have access to the university's massive centralized computational capabilities. Costs will be covered by tuition, and students will own their own computer when they graduate. Many other universities and forward-looking secondary schools will no doubt introduce comparable programs before the end of this decade. Already many U.S. school systems are making small computers available to students. The British, however, appear to have taken the lead in this area.

More than mere computer literacy and programming skills may be involved, at least for the young. The contemporary philosophy of education itself is up for grabs, as Papert points out:

> In most contemporary educational situations where children come into contact with computers the computer is used to put children through their paces, to provide exercises of an appropriate level of difficulty, to provide feedback and to dispense information: The computer programming the child. In the LOGO[3] environment the relationship is reversed. The child even at preschool ages, is in control: The child programs the computer. And in teaching the computer how to think, children embark on an exploration about how they themselves think. The experience can be heady. Thinking about thinking turns the child into an epistemologist, an experience not even shared by adults (Papert 1980:19).

COMPUTER GRAPHICS AND COMPUTER-AIDED DESIGN (CAD)

As far back as the 1950s applied mathematicians began to use computers as a tool to simulate the effect of geometrical transformations of complex shapes. One of the earliest applications of this sort was to calculate the intensity of gamma or neutron radiation in various directions from a nonspherical source, taking into account irregular shielding. It was soon realized, however, that programs to do this sort of thing could also be used to simulate views of an object from different angles and at various distances. It was a short further step to simulate a sequence of such views as they would be seen by an eye moving along a prescribed path in three-dimensional space.

The development of suitable graphic output devices, in both soft versions (displayed on CRT screens) and hard-copy (printed) versions proceeded rapidly. Thus by the early 1960s computers began to find limited applications in the commercial art and publishing business and in film animation, especially for the making of television commercials. Animated computer graphics are an increasingly common sight. The NBC peacock is but one of many familiar examples.

Application of computer graphics programs to the design of structures such as buildings and bridges was an obvious extension. At first the computer programs were simply used to help prepare complex, repetitive drawings, which require a variety of different views of the object. Soon, however, more demanding tasks were attempted, such as the graphic simulation of networks of water pipes, electrical wiring, and heating/ventilating pipes to satisfy specified constraints and requirements. Examples might be: avoid door and window spaces, provide electrical outlets at given intervals, assure adequate air movement and temperature control, and minimize the amount of wire and pipe to be used. One of the most difficult problems, still not fully solved, is that whenever any change is made in the design of the structure—for instance, the floor layout of a building—all the life-support systems must be shifted. Ideally the computer software will eventually do this sort of thing automatically, although that capability is still under development.

In mechanical engineering design, the computer can automatically carry out a variety of standard calculations, such as bending and shear stresses for a bridge or minimum thickness for a pipe. Thus the computer can test design concepts. Some specialized models are beginning to be able to calculate distortions from metal bending, heat-treating, or forging operations for certain geometries. Computers can also now simulate the effects of collisons or high-speed turns on alternate automobile designs.

One of the areas where there has been enormous progress in recent years is the development of high-level languages, whereby designers, engineers, and architects who are not professional programmers can use the computer. This development, in the early 1970s, created a commercial market for packaged CAD systems. Demand for such systems has grown extremely rapidly over the past 6 or 7 years.

The biggest users to date have been aerospace and semiconductor manufacturers. CAD systems are ideal for such tasks as redesigning a wing structure or a gas turbine blade where literally thousands of design variants may be tried before the best one is selected for production. The impact has already been dramatic. For instance, Edward Nilson of Pratt & Whitney, the largest manufacturer of jet engines, was quoted by *Fortune*: "Thanks to CAD we have gained a 5 to 1 or 6 to 1 reduction in labor and at least a 2 to 1 reduction in lead time" (October 1981). Comparable savings have already been realized by some architect/engineering firms using CAD systems:

> For example, one architect and a computerized library of preengineered components can produce all the plans for a $3 million building in one afternoon—a task that would have taken three months to do manually. Such savings reduce a firm's design costs by 30% to 40%, even when the price of the machine is capitalized. For the client, this translates to a 20% to 25% reduction in architecture fees (*Business Week*, March 15, 1982).

The most dramatic current application of CAD, however, is in the design of complex microcircuits on chips using so-called very-large-scale integration, (VLSI). Traditional methods of circuit design are very labor intensive and time consuming. CAD has been employed for this purpose since the early 1970s, and its importance grows as more and more circuit elements are packed onto a single tiny silicon wafer. The problem is simply that, as of 1980, using existing computer aids (CAD) it cost about $100 per "gate" (or circuit element) to design a VLSI chip. Thus a 64-kilobit (64K) chip costs about $6.4 million to design. But at this rate, a chip with 1 million transistors on it would cost $10 million to design (Robinson 1980). This is far more than the projected semiconductor market expects to be able to bear.

To cut design costs, CAD capabilities must be upgraded very dramatically if progress in VLSI chip technology is to continue. But CAD costs are primarily related to software, not hardware. Thus to meet the future demand for CAD, applied to VLSI circuit design alone, enormous investments in software development will be required. Indeed VLSI circuit design is one of the priority targets of research in artificial intelligence at several centers in the United States and Japan.

THE FACTORY OF THE FUTURE

Still another important aspect of the third industrial revolution is the coming widespread application of computer-aided design-computer aided manufacturing (CAD/CAM) and robots in manufacturing. The first automatic machines and tools in the mid-nineteenth century were controlled by mechanical means (cams and limit switches), which crudely combined the functions of sensor, program storage, and effector. This mechanical technology remained standard for over a century until the advent of magnetic (analog) process controllers in the 1950s. The first prototype industrial robots were also mechanically controlled, and some robots of this type are still being sold. Numerical (digital) controls followed shortly and were introduced in the first commercial robots at the same time they were first applied to machine tools around 1961. Direct numerical control (DNC) and computer numerical control (CNC) for both robots and machine tools arrived in the early 1970s. However, it is only in the last few years that the new generation of computer-controlled machine tools is substantially replacing the older types in most applications. As compared to the manual/mechanical control systems of less than two decades earlier, output per machine has risen by factor of 3 to 5. Thus although only around 4 percent of the machine tools in the United States are numerically controlled, these tools may account for as much as 50 percent of the value-added by the metalworking industries. This alone might justify the notion of a coming qualitative change in manufacturing technology.

However, computerized numerical control is applicable primarily to individual single-spindle machine tools. (The spindle is the rotating shaft to which a cutting tool is affixed.) Such machines are used in job shops and in batch production, but not in high-volume mass production. All jobs in a job shop are carried out sequentially. Mass production on the other hand requires so-called dedicated manufacturing systems. These consist of giant multiple-spindle transfer machines, generally with between 100 and 1,000 tools, mainly drills, cutting simultaneously. The spindles are clustered in groups (or stations) and the machine automatically transfers the work pieces from station to station.

The mechanical requirements of such a machine are awesome. Each of these spindles must be permanently positioned very precisely with respect to all the others. All the spindles in each group must also be exactly synchronized, so that the resulting holes are not only parallel but also drilled to the exact same depth. Drill speeds must be precisely predetermined for the same reason. The necessary simultaneity is achieved by mechanically linking all the spindles at each station, if not all stations, via elaborate gear trains to a single drive shaft.

Interestingly, high-production machines like this are themselves custom-designed and built virtually by hand. A dedicated manufactuing system capable of producing 120 engine blocks per hour more or less continusouly for 20 years—roughly 10 million engine blocks altogether—requires roughly 60,000 man-hours of engineering effort (Cross 1980).[4]

Such manufacturing systems are also remarkably inflexible, in the sense that once the transfer machine is built, it is impossible to change the design of the engine block at all. This rigidity explains the otherwise puzzling fact that U.S. automobile manufacturers were not able to convert plants making eight-cylinder engines to six-cylinder engines; nor can a plant dedicated to making conventional transmissions and universals for large rear-wheel-drive vehicles be converted to manufacturing transaxles for front-wheel-drive cars.

But computer control at the individual machine level permits fundamental changes. To begin with, it permits stand-alone CNC machines to be linked, via computer controllers and robots into integrated groups called *manufacturing cells*. Such a cell is illustrated in Figure 6-6. It operates like a transfer machine on a given part, per-

Figure 6-6. A Hypothetical Two-Robot Manufacturing Cell.

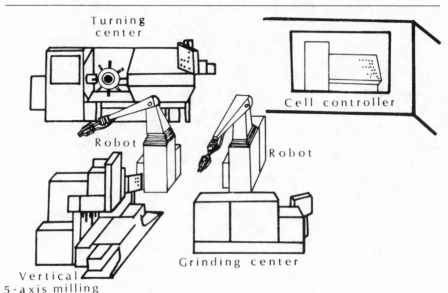

Source: Carnegie-Mellon University Robotics Institute (1983).

forming a sequence of specified cutting, forming, and manipulation operations, including initial loading and final unloading and palleting. But it can be reprogrammed to handle a different part.

The computer that supervises the cell as a whole feeds instructional programs to each of the machine tools and robots in the cell. It also coordinates materials flow between the machines, inspection of the work piece, and tool changing. In the event of equipment failure it automatically interrogates the individual machines to diagnose the problem and correct it or summon assistance.

Still higher level controllers can link different cells together, in various ways, according to the part being made. Scheduling and materials flow become the major functions of this level of process control. To achieve reasonably high levels of machine utilization in plants where a large number of different parts are made, it is necessary to exploit the geometrical similarities between families of parts, known as "group technology." Software is the limiting factor here. Special control procedures must also be designed to deal with common but abnormal situations, including special-priority orders, reworking or disposal of parts rejected by inspections, and rerouting of parts due to machine breakdowns. All of the foregoing falls under the rubric *computer-aided manufacturing*, or CAM.

An interesting question now arises: can a system of computer-controlled operations performed on stand-alone machines compete economically with a transfer machine where many operations are performed automatically in parallel (simultaneously)? At first sight the mechanically linked, high-speed parallel processor seems to have a great advantage in speed—once it is designed and built. But the advantage may be somewhat illusory. The real question is whether a single 400-spindle machine designed to carry out 100 sequential drilling, boring, or milling operations four at a time is necessarily more cost-effective than 400 single-spindle machines arranged say, in four lines of 100. The answer depends on the unit cost of single-spindle machines and on whether they can be efficiently monitored, coordinated, and programmed.

As regards comparative costs, it must be admitted that 400 single-spindle machines would also require 400 individual drive motors, motor controllers, and microprocessors—in contrast to the single transfer machine. On the other hand the individual machines would not require the complex and expensive mechanical linkage required to assure precise synchronization in the large machine. Moreover the stand-alone machines would be relatively standardized, off-the-shelf items not requiring costly custom design. They would themselves be manufactured by automated equipment (machines like themselves) in large numbers and therefore at a far lower cost per spindle than the virtually hand-made transfer machine.[5] In short, one question of great significance is whether the advent of CAM will create a cost-effective flexible alternative to the use of dedicated hard technology in mass production. Although the evidence is not yet in, it seems very likely that a flexible mass-production technology can be developed within the coming decade—in Japan, if not in the United States. Indeed it is highly probable that the availability of customized software to control the flexible manufacturing system will turn out to be the limiting factor.

It is well to recognize that computers will also bring down the cost of hard automation. One of the major costs of customized equipment, it will be recalled, is engineering design. Recall the figure of 60,000 man-hours to design a large transfer machine. Obviously the application of CAD may eventually reduce this element of hardware cost significantly, perhaps by as much as 80 percent if experience of other CAD users is a guide. This would not make the resulting equipment any more flexible than it now is, however.

The potential for manufacturing under computer control goes further still, of course. In principle, the design process can also be included. The traditional manual design technique yields part drawings (blueprints) on paper. These are then read and interpreted by manufacturing engineers and a skilled machinist, who build a prototype of the part from an appropriate semifinished material such as bar stock. In practice there may be opportunities to simplify the production sequence significantly by altering the design, to make the part easier to grip, say. More important, the design may be altered to simplify the assembly process, mainly by minimizing the number of parts, such as connectors. Once the design is fixed, the machinist or engineer then works out a sequence of machine operations and develops specifications of jigs, tools and dies, and machine settings for each step. Many of the required data are developed on the shop floor, by trial and error, using actual production machines. Eventually, after a certain amount of experience has been gained, the process sequence becomes relatively fixed.

Can computers assist in the process of optimizing the production process? Computer-aided production planning (CAPP) is already an established buzzword. But the difficulties are formidable. In the first place the output of the CAD stage consists of detailed specifications of the final shape only. The shapes of the work piece at intermediate points in the sequence—of which there may be many—are not uniquely determined by the final shape. In fact there may even be a choice of several fundamentally distinct routes to the final product, such as powder forming, forging, casting, or machining. For a given route there may be a choice of different semifinished forms to start with; for instance, bars or billets of various dimensions.

In principle there should be an optimal choice among the myriad possible sequences, but in practice this is hard to be sure of finding. Indeed manufacturers of equal sophistication commonly make different decisions.[6] What *does* appear feasible within a decade or so is for alternative production sequences to be tested by means of computer simulations. The performance of each machine tool could, in principle, be modeled, and a library of such models could be stored in the computers. It is not difficult to imagine a computer-assisted systems analogous to CAD for designing production processes. To be truly helpful the CAD system would have to have a high-level language accepting instructions from users in a close approximation of ordinary English.

One of the complicating factors in the choice of an optimum production process is that the plant may be manufacturing a number of different products simultaneously, using the same equipment. The process optimization for each product then depends on what else is being manufactured and what machines are therefore available (or not). The optimum manufacturing sequence is thus a function of the particular mix of products. There is no general once-for-all solution, even for a given product mix, because the optimal choice could also depend on the so-called shadow value or opportunity cost of turn-around time for each order. In the case of turbine blades, for instance, orders for replacement equipment normally have a much higher priority than orders for new equipment. The optimal schedule for a multiproduct operation in real time is correspondingly difficult to work out. Software for production scheduling in such an environment would also have to be modified each time a new product design appeared on the scene or else the factory control system would have to be intelligent enough to modify its own control and scheduling algorithms each time a new design was added to the menu. This would be a fairly advanced application of artificial intelligence.

The long-range implication of the substitution of computer control for manual control is that flexible automation will permit general-purpose machines to be used more in the future. Many will possess some attributes of artificial intelligence, including sensory feedback. Flexible automation will permit general-purpose shops to produce a variety of products at unit costs more nearly comparable to mass production. There is reason to believe that the unit cost of programmable flexible automation may eventually approach that commonly associated with rigid or hard automation (Ayres and Miller 1982). Instead of depending on custom-made high-speed automatic machinery to manufacture standard products such as engines, we may be able to do this just as cheaply with general-purpose off-the-shelf programmable machine tools and robot manipulators. Not being optimized for any particular task, such machines could themselves be made in fairly large numbers. Thus the capital cost of a flexible mass-production facility might eventually decline, even while its inherent flexibility increased.

An important caveat: In the early days of computers the hardware (and hardware manufacturers) defined the capabilities of the technology. Increasingly, however, it is software costs that dominate. At Bell Telephone Laboratories, nearly half of the entire research effort is now devoted to software development. According to Lewis Branscomb, IBM's vice president for research, the ratio of hardware to software costs in the computer industry itself has reversed over the past 15 years (from 4:1 to 1:4). Computers, especially mini's and micro's, are increasingly being sold on the basis of the software features they offer. Software development is probably the greatest growth industry in the U.S. economy today.[7]

In the application of computers to manufacturing, software is also becoming the critical ingredient. The cost of CAD systems is already mostly attributable to software. The next generation of sensor-based robots will, similarly, be software intensive. Moreover the coming linkage and coordination of individual numerically controlled machines and robots into manufacturing cells and fully automated factories depends largely on the development of suitable software. The shortage of software and of programmers is probably the major barrier to flexible automation in factories. However, the fact that the United States is furthest advanced in other computer applications, has a large number of trained programmers, and unquestionably leads the world in the underlying scientific disciplines—notably artificial intelligence—may be the last big comparative advantage left to this country in the coming economic shoot-out with Japan.

What flexible automation really means is that a sea-change in the manufacturing industries is in the offing. The possessor of a truly flexible production technology does not have to put a hold on product innovation in order to amortize an expensive mass-production facility. Mass production historically has meant a built-in conflict between standardization and innovation. The conflict not infrequently causes market leaders to standardize prematurely and suppress technological changes that would shorten the life of their existing investments. Flexible automation means an end to that conflict. It removes the economic restraints on product innovation.

Conventional, capital-intensive mass production of standardized products by standardized equipment is migrating to countries with cheaper labor, cheaper energy, or even cheaper capital.[8] On the other hand conventional mass production itself is on the way to becoming obsolete. Certainly it is obsolescent in the United States. Looking at it another way, *the United States can be competitive only in markets that are technologically dynamic and rapidly evolving—where com-*

petition is based more on product performance than on price. This means the ability to invent and innovate continuously is essential for economic growth and even survival in the future. Fortunately the United States still has a strong (but hardly impregnable) position in the underlying technologies of computers, microelectronics, and artificial intelligence.

THE FIFTH-GENERATION COMPUTER

All the excitement today seems to be focused on microcomputers for games, appliances, machine tools, and personal computers. But in many ways the frontier of computer technology is still at the other end of the spectrum: the supercomputer. Much insight on likely future developments in the computer field can be derived from published Japanese plans for development in this field, as outlined at the International Conference on Fifth Generation Computer Systems (FGCS) held in 1981. (NTIS 1982). These plans are highly significant for two good reasons. First, the plan emanates from Japan's formidable Ministry of International Trade and Industry (MITI), which has a nearly unbroken record of success in accomplishing the objectives set in previous plans, including the recently completed VLSI program.[9] The second reason to take the FGCS plan very seriously indeed is that the major users of supercomputers in the United States— such as Los Alamos Scientific Laboratory—are sending delegations to Japan to investigate Japanese capabilities and to inform the Japanese of projected U.S. needs (Buzbee et al. 1982). According to Professor Michael Dertouzos, of MIT's Computer Science department, if they achieve even one percent of the announced objective, the program will have been a great success. In short, the FGCS program is extremely ambitious. Achievement of the plan's objectives in full would make Japan by far the world's leader in computers—and, quite possibly, in military potential as well—by the year 2000.

Some of the key features of FGCS are listed as follows:

1. Machine speed of 10 billion numerical operations per second (10 BOPS) or 10^2 to 10^4 times faster than current Cray "supercomputers.[10]

2. Memory capacity of 1 billion bytes; retrieval rate of 1.5 billion bytes/second.

3. Speech understanding, natural language processing, and pictorial image (vision) processing. This requires a new kind of organization called distributed parallel processing architecture.

4. Multilingual translation capability.

5. Intelligence capable of learning, associating, and inferring.

The total budget for this project, over 10 years, has been projected at $500 million, excluding the salaries of government employees at MITI's Electrotechnical Laboratory.

In the hardware (computer architecture) area, another significant Japanese proposal is the so-called data-flow machine to perform large-scale parallel processing with $10^3 - 10^4$ processors and 10 GB memory (1 GB = 1 billion bytes). New kinds of solid-state devices are being investigated intensively, including Josephson junction devices, gallium arsenide and aluminum doped gallium arsenide devices (called high electron-mobility transistors or HEMTs) and high-speed silicon devices. Finally, they propose to achieve a circuit density of 10 million transistors per chip, using an intelligent VLSI–CAD system for the design and layout functions.

In the software area one objective of FGCS is a so-called knowledge-based management system capable of storing and retrieving 20,000 rules (expressed in natural language) and 10^8 data items (0.1 GB). They also propose to design and build an inference machine capable of performing $10^8 - 10^9$ logical inferences per second, (or LIPS). Each logical inference corresponds to processing 100 – 1,000 instructions.

U.S. leadership in many of the underlying technologies is being severely challenged. The Los Alamos group noted, for instance, that the Japanese already appear to be in the lead in developing high-speed gallium arsenide and high-electron mobility transistors. It also appears that the Japanese have already developed a number of parallel-processing systems.[11] Thus they already have experience and hardware on which to experiment.

The software requirements might be the hardest goal for the Japanese to achieve. One American delegation attending the FGCS conference and reporting to the National Research Council expressed skepticism: "Clearly, the goals set by the FGCS project are overly ambitious and optimistic. . . . It is doubtful that they will produce commercially successful AI [artificial intelligence] machines within 10 years." (NTIS 1981) It is important to remember, however, that the Japanese have already moved from a standing start in the early 1960s to near parity with the United States in computer technology today. Their growth rate in the computer field is double ours. Moreover MITI has identified the industry as one that is crucial for the future of Japan. Given MITI's impressive past record this is news that IBM and the rest of the U.S. computer/semiconductor industry and the U.S. government itself must take with utmost seriousness.

THE ULTIMATE INFORMATION TECHNOLOGY: GENETIC ENGINEERING

Genetic engineering is the popular term for a set of technologies based on grafting (or recombining) genetic information corresponding to specifically desired properties into the genetic material (DNA) of bacteria, plants, or animals. In the very near future, bacteria or yeast cells can be converted into biological factories to produce certain desired products of biological origin in commercial quantities, as antibiotics are now produced. Over a somewhat longer time horizon, major agricultural crop plants, domestic animals, and other commercially important species will be modified. For instance, corn, wheat, and rice are likely to be given the capability of fixing nitrogen in the same way that legumes now do.[12] In the very long run, it may even be feasible to modify human genetic material, but a great many obstacles lie in the way of any research along these lines, at least in the Western countries.

The near-term uses of recombinant DNA technology will be to manufacture various pharmaceutical products, including antibiotics and hormones, food products (such as vitamins and amino acids), animal and plant growth hormones, pesticides, animal vaccines, feed additives, various organic chemicals, various enzymes used in industry, alcohols, methane, and even hydrogen. The technique can also be used to extract certain heavy metals (such as uranium) from low-grade sources, by developing specialized bacteria that metabolize the metals and then can be harvested.

The present U.S. market value of commodities for which recombinant DNA technology seems appropriate (excluding alcohols, methane, and hydrogen) is around $10 billion. Antibiotics account for 40 percent of this figure, amino acids, vitamins, enzymes, and hormones together account for 32 percent (OTA 1982). More significant, perhaps, is the potential for producing products that cannot be synthesized by other means and cannot be extracted in large enough quantities from available natural sources. Human interferon, which may be of great therapeutic value in cancer treatment, is one major near-term target of DNA research.

Around 1980 a number of genetic engineering firms (Genentech, Genex, Cetus, Biogen, Enzo Biochem, Biologicals) were suddenly formed to commercialize research that had been going on for a number of years at Stanford, Berkeley, Harvard, MIT, and other research centers. At that time the United States appeared to be comfortably leading the world in this new field. The early lead has melted away, however, as Japanese firms have invested very heavily in research in the field (again, with help from MITI, which organized a group of fourteen Japanese companies and put up $150 million in research contacts as seed money).

When the biological products now in the offing are ready to go from the development state to the mass-production stage, Japanese firms will be ready to produce them. The reason is that they already dominate the basic production technology of fermentation. One Japanese company (Ajinomoto) has 70 percent of the world market in amino acids, and Japan is the world's largest producer of antibiotics. Counting food products (such as soy sauce) and beer, sales of fermentation products accounted for over $50 billion per year in Japan in 1982 (*Newsweek*, August 9, 1982). With this manufacturing capability, the Japanese are likely to dominate the biotechnology market.

A more interesting long-range development, now being explored, is the potential of biotechnology as a means of producing molecular-scale electronic devices, or biochips. Using knowledge and experimental techniques now being developed in DNA research, it is possible that three-dimensional information-processing structures can soon be fabricated on oriented monolayers from amino acids. Within the next two decades, moreover, it may be possible to custom design appropriate bacteria to manufacture these structures metabolically. Obviously there are enormous difficulties to be overcome. Nevertheless the implications are truly awesome because they portend a convergence of electronic and bioengineering. The ancient distinction between living and nonliving organisms may well be breached within the next half-century or less.

NOTES TO CHAPTER 6

1. It must be pointed out, however, that people who do not need to go to an office every day may move farther away, thus spreading the metropolitan area over a larger area and reducing the efficiency of other distributional functions.

2. The reasons are somewhat irrelevant to the argument, but the essential point seems to be that a relationship of trust depends in part on nonverbal communication (body language). Where the medium of communication is easily manipulated by either party, deception is also facilitated. Thus most interactions of a nonroutine nature—especially if either party is at risk—will still involve meeting in person. Managerial, marketing, diagnostic, advising, and teaching functions are in the nonroutine category.

3. LOGO is the name of a philosophy of education in a growing family of computer languages that goes with it. Characteristic features of the LOGO family of languages include procedural definitions with local variables to permit recursion. Thus in LOGO it is possible to define new commands and functions that then can be used exactly like primitive ones. LOGO is an interpretive language. This means that it can be used interactively. The modern LOGO systems have full list structure—that is to say, the language can operate on lists whose members can themselves be lists, list of lists, and so forth (from Papert 1980).

4. At a conservative $50/per hour this amounts to $3 million just for the design! The cost of an engine plant, as a whole, would be in the neighborhood of $150 million—mostly for the equipment. If transfer lines were mass produced (as engines are), the cost might be closer to $1.5 million. Stating the same facts in another way, if engines were individually designed and hand built in a job shop environment, it is estimated that the unit cost would be 100 times as great as it is.

5. Given that a complex one-of-a-kind machine probably costs 100 times more than a machine that is truly mass produced (like an automobile engine), a cost-reduction factor of 10 is probably not unreasonable for moderately large batch production.

6. For instance, Westinghouse manufactures steam turbine blades by a process largely based on swaging and forging, while GE machines the blades for its turbines.

7. In fact, worldwide software sales by U.S. suppliers are estimated at close to $7 billion for 1982. International Data Corporation projects this to rise to nearly $30 billion in 1987.

8. Capital is comparatively cheap in Japan because the savings rate is high and because the government allocates credit preferentially to export industries.

It is also cheap in some OPEC countries and, for approved projects, to clients of the World Bank.

9.   In brief, the objective of this plan was to make Japanese firms world leaders in semiconductor manufacturing. Earlier plans dealt with shipbuilding, steel, autos, and computers, among others.

10. Fujitsu, Hitachi, and NEC are all working on a "super Cray" machine to be ready in 1983–84. For example, the HAP-1 (Hitachi) to be delivered in 1983, will have a speed of 250 million operations per second (MOPS), compared to 170 MOPS for Cray-1.

11. Examples include Fujitsu M382 (two processors); Hitachi M280 H (up to four processors); NTT has an experimental system with 16 processors; Toshiba has a system with 16 parallel microprocessors; NEC has an operational parallel processor for image processing.

12. The nitrogen-fixation is carried out by certain bacteria that live in a symbiotic relationship with the roots of the plants, but the effect is the same. The challenge would be to give other plants the attributes required by nitrogen—fixing bacteria.

POLICIES TO FURTHER TECHNOLOGICAL INNOVATION

INTRODUCTION TO PART II

Most American presidents who came to power under a banner of consistent public morality and principles have conspicuously failed to deliver on either. The difficulty of applying principles consistently in politics is amply illustrated by the absurdity of trying to punish the Soviets for invading Afghanistan by hurting U.S. farmers, as did the Carter administration, or punishing the Soviets for supporting the repressive martial law regime in Poland at the expense of U.S. workers in Peoria, as did the Reagan administration.[1]

Principles are applicable to policymaking, but it is very difficult to formulate a set of principles that will not very quickly lead to conflicts and paradoxes. Moreover the difficulty is unquestionably getting worse. During the nineteenth century U.S. foreign policy was essentially constant though occasionally inconsistent. Its pillars were nonintervention, isolationism, and the Monroe Doctrine. The expedient departures from the standard policy, such as Commodore Perry's historic visit to Nagasaki "opening up" Japan to trade with the West, and the several territorial expansions at the expense of Mexico and Hawaii, were relatively exceptional, though admittedly important. The tide changed when the Spanish American War and the First World War ended U.S. isolationism de facto. Since then the principles of isolationism and nonintervention have been in more or less continuous retreat. The Monroe Doctrine too is now largely in shambles.

175

Similarly domestic civil rights policy was, for a long time, the exclusive province of the states. "States' rights" clearly superseded civil rights, at the federal level. Today the principle of states' rights is largely a dead letter.

By and large, the policy anarchy today arises from the fact that recently discarded principles that formerly governed U.S. policy have not been replaced by others of comparable generality or acceptability. One of the few general principles of recent origin—governing U.S. policy since 1948—is the "containment" policy to counter Soviet Communist expansionism and its stepchild, the alliance system—especially the North Atlantic Treaty Organization. NATO's credibility was seriously undermined in the early 1960s by de Gaulle's anti-Americanism, and its unity is now very fragile. The underlying policy itself was called into question by the opponents of the Vietnam War and by the "China Lobby" which opposed the Nixon-Kissinger rapprochement with China. For a time, active containment was superseded by a more positive policy of détente but this has been undermined by the demise of SALT II. At present relations between the two countries are bad and there are ominous signs of a disastrous reacceleration of the arms race.

One advantage of basing policy decisions on a set of broad and generally accepted principles is that it reduces the scope for choice—and error—in any given case. In foreign relations, for example, allies and enemies know what to expect and can adjust their own actions accordingly.[2]

In the domestic context, the major advantage of enunciating broad but explicit principles is to provide guidance to policyworkers for dealing with cases—and they are many—where more narrowly defined interests and objectives may clash.

One very broad principle could be summarized as follows: *Seek to change the economic incentives underlying societal behavior patterns.* Exhortation and leadership are not very effective in changing such patterns. It is generally a waste of time to exhort consumers to save their money, if savings are taxed while borrowing is tax deductible so the existing economic incentives favor spending. Similarly, it is idle to propose a new "environmental ethic" to reduce litter and recycle materials in the face of a manufacturing and distribution system based on throwaway packages. Workers will seldom be persuaded to moderate their wage demands when inflation is rampant and executives are seen to be multiplying their own incomes. Drivers will not curtail their use of cars if gasoline is very cheap. Nor will business leaders invest for the long-term future if their incomes, bonuses, and promotions are entirely based on current profitability.

The following are subsidiary principles that might be useful in guiding the making of public policy in the 1980s:

1. *Long-term planning.* Just as corporations plan for the future, so must the nation. Success will require concentration of resources (including federal R&D funds) in promising newer areas and encouraging older mature industries to do likewise. The major problem is to resist political pressure for protectionism and federal subsidies for losers "to save jobs."[3]

2. *Bringing private and public economic interests into coincidence.* It is commonly assumed in the United States—especially business groups—that, in general, what is good for business is good for the nation. This has been an article of faith for a long time. Appeals from the political left to narrower working class interest have generally failed among Americans. The twenty-first century need not be seen as a simple Marxian conflict between bloated capitalists and exploited workers. However, there is now a real and growing divergence between the public interest in creating jobs, buying power (and tax receipts) within the United States, and the interest of multinational corporations in maximizing paper profits, even if the profits in question are earned and entirely reinvested abroad and cannot be secured without substantial net exports of U.S. jobs, capital, and technology.

3. *Treating humans as assets (resources) rather than liabilities.* A more effective human resources policy would reduce political pressures for counterproductive protectionist policies. It would reduce worker and management resistance to innovation. It would also reduce worker and management resistance to innovation. It would harness the talents and abilities of many people who have something to contribute, rather than depending exclusively on direction from the "experts." A far more effective human resources policy is an essential condition for a participative democracy to function effectively.

4. *Fostering technological innovation.* Both the generation and the adoption of new technology must be encouraged. The major obstacles are in the latter area. Both passive and active forms of

resistance to change must be overcome. The key is incentives. In industry, more emphasis must be put on rewarding workers and managers for long-term results. Most important, workers (both salaried and hourly wage) must be relieved of fears of technological displacement and unemployment.

5. *Avoiding the "lifeline" obsession.* It is tempting to think that the United States must, at all costs, protect its lifelines to keep resource exporters such as Zaire (cobalt), South Africa (chrome, manganese, platinum), Saudi Arabia and Iran (petroleum). This doctrine becomes a justification for maintaining an enormous military establishment and subsidizing many unpopular regimes. It can quickly backfire, as we discovered in Iran. It also deprives us of the incentive to find technological alternatives.

One other point deserves some emphasis here. The socioeconomic system has enormous inertia. It tends to go on doing what it is doing. It contains many self-adjusting feedback loops to accommodate pressures. Thus, simplistic policies often produce results very different—even opposite—from what was intended. Minimum wage and rent control laws are good examples. The consequence of the first has been elimination (by mechanization) of many low-wage jobs and the consequence of the second has been the deterioration and loss of much of the low-cost housing stock. Policy analysis must be carried out in a very broad context. The policy cure to the economic malaise will not be found in the field of industrial policy alone. Technology policy, energy resource policy, labor policy, and many other policy dimensions must also be considered.

Policies based on slogans such as "Let the market work" inevitably tend to be simplistic and unbalanced. There is a place for income tax cuts and a place for federal budget cutting. But there is also a place for environmental regulation, social safety nets, energy-use taxes, antitrust laws, and even Keynesian pump priming. The policy problem is one of choosing wisely among a great variety of alternative approaches and instruments, balancing among competing interests in society, carefully monitoring consequences, and adjusting the policies as external circumstances change.

The next three chapters (7–9) deal with three relatively narrow government policy areas: industry and trade, human resources, energy mineral resources. These three areas all have major impacts on the rate and direction of technological change. Chapter 10 is concerned with policies at the corporate level. I have not attempted to formulate any general macro-economic policy—though this is also urgent—nor have I tried to deal with a variety of other topics (transportation, health, education) that have technological implications. Nor do I touch on foreign policy and national security policy (defense), except tangentially. To do so would require a much more comprehensive treatment.

NOTES

1. Employees of Caterpillar Corporation, which lost a $90 million export order for pipeline-laying equipment, due to the U.S. embargo of the Soviet-European natural gas pipeline project. The embargo has been a complete fiasco.
2. Of course, both of these so-called advantages can also be disadvantages in some situations, and a less principled rival power may take advantage of knee-jerk reactions. See the comments on p. 108 relative to British foreign policy.
3. Long-range planning does not necessarily have to be done in a new agency in the executive branch with power to allocate resources. In fact, the main benefit of such an activity might be to provide Congress with a rationale for *eliminating* existing tax subsidies and other market distortions.

7 INDUSTRIAL POLICY

The U.S. industrialized a long time ago during a period of ample re-source availability and scarce labor skills. We learned as a nation to substitute the one for the other. Labor-saving machines, for facto-ries, farms, and homes reached their highest development in the United States. As a result we exploited our domestic resource base so lavishly that by this century we had become a massive importer of many industrial raw materials. (In contrast, the USSR is still largely self-sufficient.) To pay for these raw material imports the U.S. econ-omy must successfully export services and at least some manufac-tured goods—in broad competition with exporters in Europe, Japan and a number of other developed countries. But the U.S. industrial establishment is having increasing difficulty in the international trade competition, not only in foreign markets but even at home, as has already been pointed out.

The industries producing basic metals (and chemicals) and stan-dardized products such as automobiles, tires, consumer electronics, cameras, watches, and so on are experiencing severe difficulties. The reason is straightforward. In these mature sectors, products are stan-dardized, competition in the marketplace is largely on the basis of price, and the lowest price requires the lowest cost of production—which is not in the United States.

Given that the technology of production in these industries is itself also largely standardized, the lowest production cost is likely to be found in a country with (1) very cheap raw materials or (2) cheap (perhaps subsidized) capital, or (3) cheap labor. In the case of Japan, raw materials have not been cheap, but capital has been relatively inexpensive to industry because of the very high savings rate and because of credit allocation by the government in favor of export industries. Labor, too, has been relatively inexpensive (at least, compared to the United States) both in terms of direct man-hour costs and, more important, in terms of labor productivity, motivation, skill, and effective deployment. Japanese unions, for instance, do not demand restrictive work rules that create phantom jobs.

Because the underlying trends are irreversible, it is totally unrealistic to believe that the loss of mass-production industries to overseas locations can be reversed by traditional macroeconomic therapy such as substantial subsidies to capital via accelerated depreciation or upgrading of plant and equipment based in the United States. In short the widely heralded notion of reindustrialization of America is a chimera, at least as regards the basic industries.

At best the migration of basic industry overseas can be slowed down somewhat. The broad implications are clear: Either the United States must find other kinds of manufactured products to export, or it must sell more services abroad, or it must cut down very sharply on imports, or all three. What manufactured products might the United States continue to produce competitively for exports? The main criterion is that the product *not be standardized*, so that production is in the flexible "batch" mode. This in turn implies a continuing rapid rate of technological innovation in both the product and its production techniques. Once standardization occurs, migration of production abroad is inevitable.[1] Continued leadership in the development of so-called high-technology products, such as military aircraft and computers, is therefore critical to the future of the U.S. economy. It is vitally important that these products be manufactured in and exported from the United States, not simply licensed to foreign producers.

Robotics and computer-aided manufacturing will contribute significantly to the ability of the United States to shift its industrial system away from concentration on low-cost mass production of standardized products to batch production of rapidly evolving products that compete on the basis of performance rather than price. It is vital, therefore, that the United States regain a leadership position in these supportive technologies. This will be a very difficult challenge in view of the strong Japanese commitment to dominate in precisely these same technologies (and for similar reasons).

The two basic problems in the realm of industrial policy are as follows:

- To facilitate the shift of capital from sunset industries (mass-produced standardized, commoditylike, products), to innovative sunrise industries (batch-produced, nonstandard, rapidly changing products).

- To foster the growth of the latter industries and their manufacturing employment, in the United States. Absent the last condition, the citizens and taxpayers of this nation have no stake in innovation or modernization.

The problems associated with conversion and recycling of workers in the declining industries will be discussed in the next chapter.

PLANNING TO ASSIST SUNRISE INDUSTRIES

This is a touchy topic, because any notion of planning has been anathema to American business leaders in the past. Since corporations themselves engage in explicit, albeit rather short-range, planning, to some proponents of industrial policy it seems utterly irrational for businessmen to oppose long-range planning at the national level, especially when many of the same business leaders openly envy the smooth government-business relationships in Japan, West Germany, and France.

The highly publicized role of Japan's Ministry of International Trade and Industry (MITI) together with the Bank of Japan in orchestrating private sector activities in Japan has often been held up as an example of "indicative" planning, in sharp contrast to the role of central planning agencies in socialist countries. In the United States at present only the Department of Defense (DoD) might play a role similar to that of MITI. But unfortunately the DoD has consistently given economic issues very low priority. For instance, the DoD has, promoted a policy of "rationalization, standardization, and interoperability" (RSI), which amounts in practice to enforced technology transfer from U.S. firms to European and Japanese firms as an inducement to those countries to increase their military preparedness. Extremely valuable advanced materials, gas turbine, aircraft, and electronics technology has virtually been given away to our allies as a result, permitting French and Japanese firms to gain market share in civilian electronics, telecommunications, and aviation markets; in competition with the United States (e.g., GAO 1982). Another problem is that national defense requirements have often been used as an excuse to justify special government support for declining industries such as specialty steel and machine tools—thus depriving them of needed incentives to change.

One of the most outspoken advocates of an active industry policy for the United States is liberal economist Lester Thurow of MIT. What Thurow proposes is a National Finance Committee consisting of recently retired business and labor leaders and government officials, with the power to provide debt capital to sunrise industries at market rates of interest (Thurow 1981). The basic argument for doing this is that our major technological competitors—Japan, France, and even Germany—are doing much the same thing through government sponsorship of investment banking. Ideally the planners would also provide financial inducements to disinvest in sunset industries, allowing them to "wither away." One of Thurow's examples is the farm equipment division of International Harvester, which he believes should be closed down to reduce financial damage to the remaining stronger firms, Deere and Caterpillar (Eads et al. 1983).

Although Thurow has argued that such a National Finance Committee need not explicitly engage in long-range planning, this seems a trifle disingenuous. If such a committee were created, it would certainly require a professional staff comparable to that of the Federal Reserve Board, and the staff would certainly build economic models requiring large databases, construct scenarios, prepare policy papers, and so on. This is certainly very close to planning.

An interesting comment on this issue has been made by George Eads:

> The debate over industrial policy is very much a continuation of the debate that has raged off and on in this country over at least the last fifty years about the feasibility and desirability of "government planning." Previously, the quickest way to end that debate was to describe the level of governmental intervention that would be required in a planned economy. (In a pinch, one could invoke the spectre of Gosplan.) But we now have an extremely high level of intervention, though we don't call it "planning." Or, to be more accurate, we have *unplanned* intervention. We got it not because we as a nation

ever made a conscious decision to have the government to assume the role of directing the details of business decisionmaking, but as an unplanned by-product of our efforts to achieve various important social goals—the cleaning up of the environment, the improvement of workplace health and safety, and yes, even the promotion of investment though various forms of direct and indirect incentives. In trying to achieve each of these ends, we have employed techniques that put the government in the position formerly held by the business decisionmaker (1983: 5–6).

The U.S. government already intervenes in the marketplace in many ways. It makes direct loans (to farmers, small businessmen, and students), buys mortgages, insures and finances exporters, subsidizes agricultural crops (including cotton, peanuts, sugar, tobacco), allocates TV channels, fixes prices (for sugar, tobacco, and interstate natural gas), stockpiles raw materials such as cobalt, ferrochrome, tungsten, and even petroleum. It sells electricity in certain regions (Tennessee and Kentucky, the Pacific Northwest), maintains navigable waterways, airports, and highways, and auctions drilling rights for oil and gas on government property or offshore. Finally, it regulates all private industry with respect to environmental impact, health and safety of employees, advertising claims and competitive behavior. It also regulates products such as food, drugs, and children's sleepwear.

In all of these and other interventions, there are opportunities for political favoritism, to say the least. One of the few safe generalizations about this maze of programs is that they are costly and economically inefficient, yet strongly defended by their beneficiaries. They raise the overall cost of government, contribute to unbalanced budgets, inflation, and high interest rates. Some programs are better than others, of course. TVA was very beneficial to a depressed region (that is no longer depressed). Loans to small businesses and farmers may have had significant social and economic payoffs on balance, though much of the money has gone to wealthy but well-connected individuals. Clearly, crop subsidies and price controls are very dubious ways of protecting certain farmers and consumers, respectively, at great expense to the rest of society.

Active government intervention by the government is largely on behalf of special interests in established but now declining sectors. Mancur Olson has pointed out why the political power and resulting economic benefits enjoyed by such groups may be far greater than their numbers or economic importance would imply (Olson 1982). The notion of creating a special-interest agency on behalf of the more dynamic, innovative sectors of the economy is not entirely un-attractive, if it could be done successfully.

Some of the proponents of a more open (or transparent) industrial policy argue that something like Thurow's National Finance Committee would make it easier to stop subsidizing "losers" (like the steel industry and conrail) and channel resources instead to future winners like the semiconductor industry. Unfortunately there is no simple formula to determine the net social benefits of any particular intervention, nor is there any political formula to keep the National Finance Committee free from political influence. Moreover, even the ad hoc back-door form of intervention is not always bad.

Lockheed, the first major bail-out, is now doing quite well, having long since paid off its debt to the government. In the case of Penn Central there really was not much choice, since rail services are clearly indispensable in the area served by the system. The only real question was whether the railroad should remain in private owner-ship or not. It may be sold back to its employees. (Outside the United States virtually all railroads are nationalized.[2]) Chrysler may not be completely out of trouble, but it has paid off its loan, plus a premium, and it has also renegotiated its contract with the workers. Its survival would not have been possible at all without assistance. Had Chrysler collapsed, the resulting domino effect would have also bankrupted many other reasonably well-managed businesses (sup-pliers and dealers). A *Business Week* editorial cogently pointed out: "If the haze of unreal expectations that has gathered around high-tech industry demonstrates anything, it is that nobody—especially government bureaucrats—possess the wisdom and foresight to decide which enterprises should live and which should die" (March 28, 1983).

THE TRAP OF CONVENTIONAL PROTECTIONISM

There is one common element in the policy proposals of virtually all declining industries, and organized labor. It is the demand for protec-tion against unfair foreign competition. Protectionism of the old-fashioned kind, high tarriffs, is no longer widely practiced by the advanced industrial nations, because of the General Agreement on Tariffs and Trade (GATT). However nontariff barriers to trade are another matter entirely.

In the United States the most common type of protection is to negotiate bilateral quotas for specific exporting countries, or multilateral quotas for specific products. The United States has a very complex system of sugar quotas, for instance, to protect domestic sugar producers. Important quotas exist for textile fibers, shoes, and a host of other products. Japan is currently limiting its exports of automobiles to the United States under a three-year quota, due to expire in the spring of 1984. It may be extended for another year if current negotiations are successful.

Other countries use different techniques to foster and protect domestic industries and avoid the rigors of free trade. Some of these methods, such as direct export subsidies, are prohibited under GATT rules but hard to detect and enforce. Others are seemingly impossible to control. For instance, nationalized European steel manufacturers have access to cheap capital from the public treasury. In Japan the banking system allocates credit preferentially to export industries (at the expense of domestic consumers). European countries also finance sales to foreign customers—including Eastern Europe—by providing subsidized below market long-term loans, again from the public treasury. The U.S. Export–Import Bank has a similar function, of course, though it operates on a much smaller scale and cannot subsidize interest rates to the same degree. Japan is particularly adept at excluding foreign-made goods by imposing a complex nightmare of safety tests and other requirements. Finally most countries manipulate their currencies to keep currency exchange rates as favorable as possible. In Europe, where capital supposedly moves freely, this mostly takes the form of central banks buying up dollars, thus bidding up their price. The reserve dollars can then be used later to finance exports, or to buy gold, if the price of gold falls. The Bank of Japan keeps the local interest rates below world levels but denies access to foreign borrowers, thus keeping capital costs low for Japanese manufacturers and also keeping the Japanese yen undervalued by something like 15 percent. This manipulation is possible because the yen is not a reserve currency, so any yen holders must ultimately sell them back to the Bank of Japan—in effect a monopolist consumer. This device strongly favors Japanese exporters in foreign markets, especially the United States. In recessionary periods Japan typically increases its exports, thus keeping its own workers employed at the expense of workers in other countries.

Obviously some of these clever devices are not available to the United States precisely because the U.S. government plays a smaller role in business than do the governments of most other countries, and the overall result is distinctly one-sided. The fact that the dollar is the chief reserve currency of the world does not help matters. It is thus understandable that U.S. manufacturers and workers seek relief from their own government. The problem is that some of the proposed cures would make the problem even worse.

The AFL–CIO approach (projecting somewhat from current union policy) would be to ask for such things as

- Tougher "antidumping" rules
- "Domestic content" rules for imported cars (or higher tariffs, quotas, etc.)
- Advance notice of management intention to close plants, eventually making this a matter for collective bargaining
- Increased security for existing employees (but without significant relaxation of work rules that would give management flexibility to adjust).
- Mandatory severance pay for displaced workers
- Controls on capital exports
- Restrictions on "outsourcing" (purchasing components from nonunion or foreign supplies)

The general thrust of this program is highly undesirable. It seeks to perpetuate an overly cozy alliance of big labor and big business with government playing a more and more explicit role. It seeks to protect both established U.S. industry and organized labor from competition from more efficient producers and more efficient workers at the sole expense of the consumer and taxpayer. It will raise producers' costs without improving their performance. It does nothing to encourage innovation or to discourage featherbedding.

Business generally seems to oppose explicit subsidies or other direct assistance for the declining smokestack industries, although these industries already pay effective tax rates that are much lower than those paid by healthy, growing firms. But, according to a Harris poll a surprisingly large number of businessmen, both inside and outside of the affected industries, seem to favor assisting these industries by giving them access to the tax-exempt bond market. There are two obvious objections to this crude scheme. In the first place a tax subsidy is still a subsidy (on top of existing tax subsidies) and is still a cost to the public treasury. It would favor the inefficient steel industry and other sick industries—which already pay virtually no tax—and, by soaking up more of the small supply of available savings it would make capital scarcer and raise financing costs for everybody else. There is no compelling social reason to do this.

The second objection is practical: If a steel company could sell tax-exempt revenue bonds, what is to stop the firm from using the money to buy an oil company or an electronics company instead of investing it in modernization of antiquated steel mills?

A case can be made for some use of tax-exempt bonds in certain special circumstances but emphatically not to infuse new capital into existing stockholder-owned firms. The only circumstances under which this mechanism is worth considering are that a still viable, but barely profitable, facility is spun off to a group of its employees and local community institutional investors under the employee stock ownership and purchase (ESOP) plan. National Steel Corporation is in the process of doing something like this with its Weirton, West Virginia plant. It could be good public policy to encourage such spin-offs, before the plants in question become candidates for closure, by giving each new-hatched entity a one-time access to the tax-exempt bond markets, not so much to benefit the industry as to give the workers and their communities a decent chance to determine their own destiny.

A strategy of encouraging ESOP buy-outs is surely not enough to bring the interests of multinational corporations into consonance with those of the citizens and workers of the United States. The strategy of restraining imports, a version of protectionism, is not a primary recommendation of this book. It would be clearly preferable, for instance, to persuade other countries to lower their existing barriers against U.S. agricultural products and to persuade our trading partners, the Japanese in particular, to stop rigging currency exchange rates against the dollar. We must also do whatever we can with fiscal and monetary policy to become more productive and more effective competitors in international markets—short of exporting large amounts of employment.

Although conventional protectionism (tariff, quotas, subsidies) has undesirable aspects, capital exports may have to be marginally discouraged in the future. It may be necessary to (1) reduce the dollar's present role as an international reserve currency and (2) to discourage future capital exports by introducing a differential between allowable depreciation rates outside of the United States as compared to domestic depreciation rates. The purpose of reducing the dollar's role as a reserve currency would be, primarily, to increase the role played by other major trading nations, especially Japan. If all major international transactions were denominated in terms of a basket of major currencies, such as the special drawing rights (SDRs) backed up by reserves held by the International Monetary Fund (IMF), the burden now placed on the dollar would be reduced. It would be more difficult for the central banks of nonreserve countries to manipulate exchange rates in order to keep their own currencies (and exports) cheap, at the expense of the U.S. economy.

The argument for placing barriers in the way of future capital exports is not easy to defend on the basis of existing historical evidence.[3] It is, at best, a second-best solution, if truly free trade is unattainable—which seems likely. The argument for it is strictly pragmatic: Assuming current trends continue, the only viable export industry the United States may have left in the 1990s will be military hardware. It is preferable to restrain imports—even at the risk of encouraging isolationism—than to have to rely on selling armaments around the world in order to buy petroleum from the Middle East and consumer products from the Far East. Nevertheless, the standard menu of protectionist policies (noted above) appears to be self-defeating.

Many people will object that restricting capital exports may not work in the manner intended. Clearly, some capital exports in the past have secured footholds for U.S. firms in foreign markets and paved the way for exports from U.S. parents to and through their foreign subsidiaries. Obviously this is a reciprocal relationship. Foreign subsidiaries, having successfully penetrated their local markets, can then later turn around and ship their products back to the U.S. parent, displacing jobs in the United States. They can do this because they typically have the same technology as the parent, while benefiting from lower labor costs and newer plants.

Moreover governments of the more sophisticated developing countries, such as Brazil and Mexico, often make entry permission for foreign firms conditional to negotiated commitments to export a significant portion of their output, either to third countries or back to the parent country (Malmgren and Baranson 1981). To cite one example, IBM must export three computers from Brazil for every one it sells in the Brazilian market. On the other hand, there is absolutely no regulation or control on this sort of behavior from the U.S. side. Obviously this situation is a far cry from free trade as the concept is traditionally understood. Inhibiting future capital exports from the United States would not magically solve the problem, but it should result in a somewhat higher level of future investment within this country. This would presumably increase the productivity of American plants, as compared with plants belonging to foreign subsidiaries, thus keeping more American workers employed than otherwise.

There is an objection to this idea, namely that it might hobble some U.S. based firms in their efforts to compete in international markets. The semiconductor firms, which do much of their assembly at overseas "export processing zones" (EPZs) fall into this category. These firms should perhaps get other kinds of technological assistance but massive export of jobs is not the right way to solve their problem.

SHOULD WE CURB TECHNOLOGY TRANSFER?

In connection with the same set of problems (the export of jobs) a related factor is the transfer of leading-edge technology to offshore producers, whether they be subsidiaries of U.S. based firms, or not. Technology transfers resulting in overseas production that displaces domestic production are clearly damaging to American workers, their communities, and to the taxpayers in general. For example Boretsky (1973, 1977) has compared three alternatives for a typical high-tech product: production in the United States and export to foreign markets; production by a foreign subsidiary; and production under license by a competitor (see Tables 7-1 and 7-2).[4] Mansfield et al. (1983) among others, pointed out that technology transfers from U.S. firms to subsidiaries in developed countries are occurring more quickly in recent years. Based on a sample of 37 innovations analyzed in depth, during the period 1960-1968 only 27 percent of the technologies transferred to such subsidiaries were less than 5 years old. However, during the more recent period (1969-1978), about 75 percent of the transfers involved new technologies.

This phenomenon is very discouraging to American workers. It is unfortunate that some technologies, especially in electronics, devel-

Table 7-1. Alternative Production Scenarios.[a]

	Standard-Case Production in the United States and Export from the U.S.	Alternative 1 Production in a Foreign Subsidiary of U.S. Based Firm	Alternative 2 License to a Foreign Firm
Dollar value of sales of U.S. firm	100	105	4
Earnings of U.S. firm	100	330	160
Employment in U.S.	100	15	4
Tax revenues to U.S. government	100	45	25
U.S. balance of payments	100	10	4

a. The situation depicted applies only in the first few years. It actually understates the negative impacts in later years by a large margin. As the foreign subsidiary (or independent licensee) gains production experience and capability, it will begin to export the product back to the United States, thus displacing additional U.S. workers and further reducing tax revenues of federal, state, and local governments. Thus, if a U.S. based firm initially employs 100 workers producing for export, the transfer of production technology to a foreign subsidiary, joint-venture, or licensee could not only eliminate the jobs of these workers but also a further group of jobs of workers currently producing for the U.S. market. The fact that U.S. consumers benefit from lower prices in this way *may or may not* compensate for the losses of income and tax base.

oped in the United States have been transferred abroad by U.S. firms in recent decades almost as fast as they reached the marketplace. It is almost a national scandal that the consumer electronics industry, based entirely on technology originating in the United States, is now virtually centered in Japan and other Far Eastern countries. The hottest consumer electronics product of 1980s, the video cassette recorder (VCR) is not even manufactured by any U.S. based firm. Unfortunately too, the U.S. government itself is partly to blame for this absurdity. Bell Telephone Laboratories was forced into a consent agreement by the Anti-Trust Division of the Justice Department to transfer and license its transistor technology to all comers—for a

Table 7-2. Value Added to Electronics by Processing in Export Processing Zones in 1979 (*thousands of U.S. $*).

| Source Country | Tariff Schedule | | Total Value of Merchandise |
	TS806.30	TS807	
Hong Kong	73	42,222	84,263
South Korea	711	97,119	245,138
Malaysia	21,243	218,858	616,783
Mexico	5,001	39,340	112,481
Philippines	782	62,335	200,785
Singapore	35	146,753	398,945
Taiwan	13,628	43,865	101,018
Thailand	4	12,142	46,714
Other	—	—	—

Source: U.S. Department of Commerce.

mere $25,000 initial fee. Thirty-five companies (10 foreign) availed themselves of the opportunity. Many of these firms are now major exporters of electronic goods back to the United States.

The Defense and State Departments, in their zeal to modernize and standardize the weapons of NATO countries and other allies, have also forced U.S. manufacturers to transfer basic technology with civilian applications to foreign competitors under military "co-production" programs. Japan's growing aircraft industry has already been heavily subsidized by the United States since World War II via military co-production programs. Now the Japanese government is assisting the Japanese aircraft industry to turn its attention to civil markets, an extremely ominous portent for the U.S. civil aircraft industry, already under pressure by the government-sponsored, pan-European consortium, Airbus Industrie, Ltd. (GAO 1982).

Machinery already under the Export Administration Act exists to control the export of advanced U.S. technology embodied in products to potential enemies—meaning the Soviet bloc. The U.S. Department of Commerce maintains a list of some 200,000 products with "military potential," for which export licenses are required to guard against transshipment to the Soviet bloc. Products on the list accounted for $20 billion in export sales last year. Under this legislation the government may arbitrarily restrict exports either for reasons of "national security" or for reasons of "foreign policy." Under the law foreign subsidiaries and licensees of U.S. firms are subject to the same rules as the U.S. parents.

In practice, this law seems to be ineffective in terms of its purpose, though it imposes serious costs on U.S. firms. It is unquestionably a source of serious friction with friendly countries, as was exemplified in the case of President Reagan's attempt to embargo the Soviet natural gas pipeline deal with Western Europe. There is good reason to believe the cumbersome licensing requirements of the Export Administration Act are severely hampering U.S. exports in competition with Japan and European countries. As the law now stands, if a similar product is available from a foreign competitor there is, ipso facto, a good case for obtaining an export permit. The Reagan administration, with its hard-line anti-Soviet policy, wants to tighten the definition of foreign availability and to require negotiations with foreign governments to persuade them to control exports of such products. This would put an indefinite hold on export permits for U.S. based high-technology firms and provide a golden opportunity for their foreign competitors.

A 1976 report to the Pentagon by J. Fred Bucy, president of Texas Instruments, seems to make more sense. Bucy proposed controls on *technology* exports, rather than on product exports. No doubt some product export controls would still be necessary, but they could be largely limited to weapon systems. The red tape involved on the government side could be cut by 50 to 75 percent, according to some estimates, and the existing difficulties for U.S. exporters would be largely eliminated.

The approach to national security-related technology export controls recommended by Bucy should be adopted by our government. It should be broadened to include advanced nonmilitary technology also. What we must have is an American version of Japanese Ministry of International Trade and Industry (MITI), with explicit responsibilities in the area of technology transfer. For purposes of discussion, call it the Commission on Technology and Trade, or COTT. It should probably be an independent agency with formal liaison with and the Departments of Treasury, Commerce, and State, NASA, and the National Science Foundation (NSF). It would have the following four basic mandates, and the necessary legal powers to enforce its policies:

1. To initiate, finance, and coordinate joint ventures among U.S. firms of conducting applied research of national economic significance, and to coordinate the dissemination of results among U.S. firms. Such projects must be free of antitrust impediments (especially the threat of private treble-damage lawsuits, discussed later).

2. To sponsor the creation of trading companies[5] with exclusive rights to import or export certain categories of goods, such as food, natural resources, weapons, and advanced scientific equipment (including supercomputers).

3. To designate and protect new products developed and introduced in the United States from foreign competition for a statutory period of 3 to 5 years, depending on the size of the investment involved.

4. To regulate "naked" technology exports of licenses, know-how, and turnkey plants. In general there should be no restriction on the transfer of old technology—more than 5 years old—but to export a new technology, the firm would have to show a net benefit to the nation. For instance, it might be justifiable to use a license as leverage to gain access to an equally valuable foreign technology. COTT would not normally interfere in licensing deals to foreign firms in competitive markets. It would normally be involved in negotiations with foreign governments (such as the Eastern Block) or with foreign cartels subject to similar government restraints, as in Japan. Co-production arrangements negotiated by DoD should be subject to approval and potential veto by COTT.

Some serious objections to this idea will be raised—notably that COTT necessarily involves bureaucratic mechanisms. However, the stakes are now so high that no good alternative to some such controls exists. The technology export controls would apply mainly in the leading-edge electronics, computer, pharmaceutical, and aerospace and associated capital goods sectors. Technology transfers to developing countries would in general be unaffected. The purpose of controls would not be to prevent trade but to assure reasonable trading terms on behalf of United States-based firms and their employees when dealing with foreign governments or cartels.

THE LAW AWRY: PRODUCT LIABILITY
AND ANTITRUST

As Chapter 3 pointed out, innovation is not something that occurs spontaneously. In a meaningful sense it is usually induced by a clearly perceived opportunity or a social need such as scarcity of some critical resource. But a social need or scarcity alone is not nearly enough. Innovation is risky and one criterion for innovation would seem to be the prospect of monetary or psychological reward. If the penalty for failure in a risky enterprise is too severe—debtor's prison, for instance—risk-taking will be discouraged.

Fortunately this particular penalty is no longer applicable, but another form of financial risk to innovators has recently grown in importance: tort liability. A number of legal innovations in the past two decades have shifted the burden of proof from plaintiffs to defendants in cases of personal injury or death attributable to product defects. Courts are increasingly likely to hold for plaintiffs if there is any evidence that defendants failed to utilize the "best" (or "safest") available technology. If the product is later improved, courts frequently admit this to be de facto evidence that a better design was possible. The underlying legal theory seems to be that this test will provide effective inducement to potential defendants (virtually all manufacturers) to introduce improvements as soon as possible, to minimize their potential liabilities for accidents arising out of inherent design weaknesses.

In practice, however, the use of this criterion in tort liability cases seems to be having an extremely perverse effect. The problem is that whenever a manufacturer does introduce an improvement, the question is likely to arise, sooner or later, in some product liability case: Why was the improvement not made sooner? Was the manufacturer negligent? If the jury can be persuaded of the affirmative, it may conceivably award millions (or even billions) of dollars in damages to victims of accidents or occupational hazards that "might" have been avoided. To make matters worse, if anything can go wrong directly as a result of the improvement, no matter what the overall balance of social costs and benefits, it also counts as a potential cost to the innovation. In practice, therefore, each technological improvement to a mature product, such as an automobile, for instance, becomes a potential legal time bomb. More radical innovations are being scuttled by corporate lawyers even before they see light of day. An example is the General Motors "Lean Machine," an experimental 400-pound, 3-wheeled, 1-passenger commuter vehicle with the appearance of a teardrop, the acceleration and fuel economy of a motorcycle, the road-hugging and cornering characteristics of a racing car, and a high degree of protection for the driver. A sure winner, one might naively suppose. But no. Some accidents are inevitable, no matter how well designed the vehicle, and the potential legal liabilities are unlimited. So the "Lean Machine" will not be built, at least by General Motors.

The present state of affairs is clearly untenable. A fundamental reform of some sort is needed. Ideally any product improvement that is socially beneficial *on balance* should be introduced as expeditiously as possible. The problem will ultimately have to be addressed by specialists, including insurance companies and members of the bar. The most rational policy would be some sort of uniform product liability law that includes the following elements.

- A strict limit should be placed on the maximum liability for accident or death resulting from a design defect if it can be shown that the product, when introduced, met all applicable[6] health and safety standards set by relevant government agencies.

- In the case of products for which no applicable standards had been set by government agencies, there should be no presumption of negligence or correctable fault based on the fact that the product was later improved. Only *direct* evidence that some alternative design without the objectionable feature was *considered and rejected at the time* would constitute admissible evidence of negligence.

- Courts should not permit defendants in tort litigation to unilaterally co-opt secondary defendants unless the judge is convinced by a showing of evidence that there is a strong probability of an inequitable allocation of burden unless the cases are linked.

- The federal government should mandate no-fault product liability insurance.

With regard to antitrust laws, there are two problems that must be dealt with by the Congress. First, under the present form of the law, firms based in the United States can be forced by the Justice Department to license valuable, proprietary technology to all comers—

including foreign firms—in the name of increased competition. Thus Bell Telephone Laboratories was forced to license the transistor and virtually all of its other major developments to all comers, for negligible monetary considerations. Virtually all the major firms based in this country have been forced to do likewise at various times. The antitrust law must be altered to give recognition to the fact that without technological superiority, firms based in the United States *can no longer compete* in many industries.

The second problem is that under present laws it is difficult for several firms in the same industry to collaborate in R&D consortia, no matter how the resulting technology is to be shared. In this case, the chief problem seems to be a feature of the law added a few years ago that makes it possible for proven victims of anticompetitive behavior to collect treble damages, as a deterrent. The impact has been to induce a massive number of speculative lawsuits and to create a whole new class of lawyers specializing in private antitrust litigation, much of it directed at the most prominent and successful firms, like IBM. Congress should eliminate the treble damage clause, except in very restricted circumstances.

These recommendations have been made in the past and have generally died in congressional committee. That is, they have been rejected as draft legislation before reaching the attention of Congress as a whole. The reasons are fairly clear: while manufacturers are generally worried about the cost of product liability, and private antitrust litigation they have either not considered the adverse impact on innovation or they have not regarded a slowdown in the pace of technological innovation as being a serious competitive disadvantage to them. When Ford and Chrysler only worried about General Motors and vice versa, all three firms were happy to minimize the destabilizing effects of product innovation.

In addition, both trial lawyers and insurance companies have large vested interests in the system as it now functions. Trial lawyers righteously resist any legislative limitations on the rights of plaintiffs, but they are primarily defending their own incomes. Most tort and antitrust cases are speculative: that is, the attorney gets paid only if the plaintiff wins. In practice, therefore, plaintiff's lawyers collect a substantial portion of all judgments; figures are scarce, but the average seems to be in the neighborhood of one-third. Winners indirectly pay the legal costs of losers too of course, since the latter are overhead expenses to law firms.[7]

In the case of insurance companies, under the present system at least 40 percent of all premium income goes to operating costs and overhead (including lawyers' fees). Insurance companies dislike no-fault insurance because they generally collect lower premiums and pay out more to policy owners. When all the transaction costs are considered, on both sides, it appears that for every dollar received net (after lawyer fees) by a plaintiff, at least $2.50 had to be collected, one way or another, from the defendants. In other words 60 percent of the money is siphoned off into the pockets of lawyers and insurance firms. In the case of asbestos-related litigation, more than 60 cents out of each dollar of awards recovered has gone into lawyers' fees and court costs. This explains why the legal profession has been so lucrative in recent years (and, incidentally, why it attracts so many bright, ambitious young people). If a government agency or a charitable foundation dedicated to distributing program funds to needy persons kept 60 percent of the funds for overhead purposes, it would be considered a major scandal. Yet that is the precise situation in tort liability and private antitrust cases.

The public tends to consider all this a harmless game involving big corporations, insurance companies, and law firms. But the transaction costs are immense, and they all get tacked onto the prices of goods and services and passed on to consumers. Bad as this is, the resulting drag on technological innovation may be an even greater cost to our society—one we cannot afford. For this reason it is very important for the Congress to rationalize the product liability system and the antitrust laws to ensure that they do not inhibit technological innovation, without which the United States cannot continue to have an internationally competitive economy.

MERGERS AND ACQUISITIONS

It has been pointed out very cogently, especially by Reich (1983), that top managers in many firms today are essentially "paper entrepreneurs" who focus their efforts primarily on reshuffling existing assets rather than creating new ones. The most obvious indication of this is the continuing obsession with growth by merger and acquisition. Notwithstanding theoretical arguments to the effect that mergers create economic benefits via synergy and scale—otherwise why would a rational manager propose one?—there is disquieting evidence that many if not most acquisitions are really exercises in personal aggrandizement by ambitious chief executives. The recent Bendix–Marietta fiasco seems to be a case in point. Probably more to the point, outright acquisitions of innovative smaller firms by large conglomerates "managed by the numbers" are all too likely to shut off the flow of innovations by imposing inappropriate and counterproductive rules on the subsidiary. Minority investments in such firms to provide additional capital—as recently exemplified by IBM's purchase of stock in Intel and ROLM—are a very different matter. Such investments should not be discouraged by overzealous antitrust enforcement.

As mentioned in Chapter 2, I essentially agree with Reich's indictment. The question is, What can and should be done to discourage mergers? Direct antitrust regulation by the government is already excessive, and I do not favor any broadening of the role of the Justice Department. One relatively small change in the law, however, could make a significant difference. I propose a simple "disclosure" law, to be enforced by the SEC.

The rule would state that the stockholders of a firm must be informed and approve, in advance, any large purchase of stock in another firm proposed by management. It would also require stockholder approval of any sale of controlling interest and of any "golden parachutes" for top management. The SEC would be responsible for approving the prospectus for any such proposal from the standpoint of accuracy and completeness. The change would not prevent negotiations between "consenting" merger partners, who could then jointly present a merger prospectus to their respective stockholders. It would, of course, make "raids" much more difficult. The disclosure requirement would, one hopes, force management to justify proposed mergers in economic terms much better than has typically been the case in the past. Otherwise, the operation of the free market would not be interfered with.

NOTES TO CHAPTER 7

1. A simple model to explain trade between an innovative high-income country ("North") and a noninnovative low-income country ("South") has been formulated and analyzed by Krugman (1979). The model shows that North

and South can achieve an equilibrium relationship with each other and rising incomes for both. In equilibrium North develops and produces new products, while standard (old) products migrate to the South. But North cannot maintain its high-income status *unless it continues to generate new products.*

2. Japan is the only significant exception. About half of Japan's railway tonnage is carried by privately owned railroads.

3. Advocates of unfettered foreign investment note, for instance, that the net *return* from U.S. corporate foreign investments in 1980 was $26 billion, as compared to net investment *outflow* in the same year of only $3.8 billion. Moreover the surplus of returns over outflow has risen steadily and sharply, from $3.5 billion in 1970 to $22.2 billion in 1980. Of course this means only that foreign investments made in the 1950s and 1960s are now paying off handsomely—not that new foreign investments will necessarily do so in the future.

    Another set of arguments in favor of foreign investment centers on the issue of job exports. Sample survey evidence compiled by Business International Corporation since 1972 suggests that foreign-investment-oriented manufacturing firms have increased U.S. employment considerably faster than other U.S. manufacturing firms. Survey data also suggest that most U.S. firms investing abroad have not done so, in the past, to replace U.S. production. They have invested, primarily, to develop foreign markets for U.S. products, as indicated by a strongly favorable balance of exports over imports. Imports by U.S. manufacturing companies from their own foreign affiliates averaged 0.5–0.6 percent in 1960, 2.1–2.3 percent in 1970 and 3.9 percent in 1980. (Variations reflect different samples of responding firms.) On the other hand the data do show that imports from foreign affiliates are apparently increasing quite fast, though they are still relatively small. Nor does it accord with the pure theory of free trade (discussed in Chapter 3). The survey data may be somewhat biased. There were almost no respondents in the latest Business International survey from consumer electronics, semiconductor manufacturing, computers, or aircraft industries.

4. A somewhat related problem, is the closing of U.S. based plants serving the U.S. market, and the shifting of production to "export processing zones" or EPZ's in places like Hong Kong, Taiwan, Singapore and Malaysia. The magnitude of the problem is shown in Table 7-2. In 1979 $700 million worth of assembly labor was done in EPZ's, rather than in the United States. It must be pointed out that the assembly workers in EPZ's typically earn less than $3/hour and that the U.S. semi-conductor industry claims it could not compete with Japanese prices if it had to pay U.S. wages.

5. The proposed trading companies would be quite different from the tax-sheltered DISCs (Domestic International Sales Corporations), which have recently been declared to be "unfair" export subsidies under GATT rules.

6. The question of applicability remains a potentially thorny one.
7. The number of speculative lawsuits would be sharply reduced if law firms were not permitted to charge the costs of unsuccessful litigation to office overhead but were forced to treat them as investments (out of after tax income).

8 HUMAN RESOURCE POLICY

TECHNOLOGICAL DISPLACEMENT

It must never be forgotten that the ultimate purpose of trying to renovate the tired and rusty U.S. economic engine is to improve the economic well-being of all Americans.[1] Unless the benefits are shared widely there will be tremendous resistance to change and our international leadership role will likely be lost before the end of this century, if not sooner. Perhaps that will ultimately happen anyway, but our national future is still in our own hands to mold. The first requisite for a successful adjustment is that participation in the new opportunities be broad, not narrow.

The most critical question in regard to human resource policy is whether the new technologies will automatically create enough good new jobs to absorb the workers inevitably displaced by office automation, computers, CAD/CAM, and robotics. If the answer is yes, then no government policy initiatives are needed. This is, more or less, the Reagan administration position. If the answer is no, then we are wasting precious time.

The first question was addressed fairly conclusively by Herbert Simon (1977). Simon's argument is rather technical, but the basic idea is straightforward. Any improvement in productivity, whether due to improved labor skills, management organization, or smarter machines, will make it possible to produce the same output with less (labor or capital) input. Obviously it follows that using the same labor force more efficiently will produce more output, hence more revenue. If the revenue is recycled through the economy as wages and profits, both workers and owners[2] will be better off. In short, as Simon says, "The main long run effect of increasing productivity is to increase real wages—a conclusion that is historically true and analytically demonstrable" (p. 159). Of course, he is talking about aggregate wages. Simon also points out that "any level of technology and productivity is compatible with any level of employment, including full employment." So, if there is a problem, it is distributional. More precisely, the problem is to create effective redistributional mechanisms and to destroy institutional barriers to mobility.

It is painfully obvious that the needed employment redistributional mechanisms do not exist or are not working well in practice. All sorts of institutional arrangements have been created over the decades to prevent the labor market from working as it should. The explicit goal of organized labor is to use monopoly power to raise the wages of unionized workers by controlling and limiting the labor supply. This strategy has succeeded to some degree, and the unionized workers who have jobs are sometimes paid twice as much, or more, than workers with comparable skills would get in an open market.[3] Union leaders assert this as evidence that nonunion members are exploited and underpaid, but it is only evidence of the consequences of restricting free entry to certain labor markets. Such restriction raises costs in the unionized sector, and, by an iron law of economics, reduces demand for the product or service of that sector. If the product is one that can be imported, like steel or automobiles, it also results in loss of jobs to foreign countries.

Other restrictive mechanisms, including professional licensing at the high end of the scale and minimum-wage laws at the low end, also tend to raise real wages for the employed, by reducing job opportunities for the rest of the potential work force. The ideal way of maximizing employment opportunities for all would be to eliminate all artificial restrictions on competition in the labor market. Among other things this would mean outlawing collective bargaining and strikes. Such an event seems highly unlikely to occur and would probably be very undesirable from other perspectives.[4]

This leads us to consider possible second-best solutions, namely ways in which the government might intervene to compensate for institutionalized labor-market distortions and rigidities.

Ending the minimum-wage law should be one goal of policy. It has been misguidedly supported in the past by the very people who are most adversely affected by it. The true beneficiaries of the minimum-wage laws are the unions, whose wages are generally far higher. But this reform would create only a modest number of unskilled jobs suitable for teenagers or part-timers. A much more serious problem exists for men in the prime of life, many of whom are now finding themselves unemployed because of the decline of mass production and basic industries in the United States. Even if these industries recover, in the short run, very few of the lost jobs will be recreated, because there are now many semiskilled industrial jobs that can be done by robots or other machines that are both more reliable and cheaper (per hour) than human workers. A few years from now the displacement problem may also become severe for many older clerical and other office workers, primarily female.

We are indeed very fortunate that the number of persons entering the labor force annually will be declining for the next 20 years, so that natural attrition will, on the average, ameliorate the displacement problem significantly. In fact, with even modest economic growth, we can foresee a labor shortage in the United States in the 1990s.

Nevertheless, for experienced workers there is a potential of displacement, particularly in the northeast and the Great Lakes states, where manufacturing industry is most heavily concentrated. Roughly half of all employment in the metal-working industries is concentrated in only five states: Michigan, Ohio, Indiana, Illinois, and Wisconsin. Since these are the states where plants are the oldest and wages highest, any further shrinkage or consolidation of the industry is likely to be concentrated in these same states, where unemployment is already well above the national average.

ASSISTANCE IN JOB SEARCH AND MOBILITY

To the extent that unemployment is a transition problem of matching available jobs with available workers, it seems to make sense to provide assistance in providing both job seekers and employers with information about each other. It also might make sense to help finance unemployed workers to move their households to more favorable locations.

With regard to job search assistance, there seems little doubt that this is a cost-effective antifriction activity, at least in relatively normal periods. A variety of mechanisms are available, including the federal-state employment service (with 2,600 offices throughout the United States), and ad hoc job clubs, and job fairs. The employment service handles fairly large numbers of cases. In 1979 15.5 million job seekers and 9.5 million job openings were registered, with 4.5 million placements (Bendick 1982).

On the other hand the system is less effective as the unemployment rises, since staff members become increasingly required for processing unemployment insurance claims. Also, despite the impressive statistics, vacancies listed with the job service tend to be poorly paid entry-level positions in the local area. Distant jobs are seldom listed. A survey of 30 studies of job-seeking behavior found that less than 7 percent of successful job placements were accounted for by the service. Lack of counseling advice or assistance in interviewing skills, résumé preparation, and the like seems to be the major drawbacks.

A more effective model seems to be the Canadian Manpower Consultative Service (MCS), which takes a more active role, especially in situations such as mass layoffs from a major plant closing (as in Flint, Youngstown, Fort Wayne), where conventional institutional resources become overloaded. The MCS plays a coordinating role in the mobilization of local resources (employers, unions, government). It also provides a case officer and a small amount of funding, which must be matched locally.

Mobility assistance has been tried out on a small scale several times over the last 15 years, but with limited success. Most such programs provide faily limited cash benefits—no more than a few hundred dollars. It appears, however, that this level of funding is not critical to the relocation decision of young, single, unencumbered individuals, most of whom would have been willing to move anyway. On the other hand there is evidence that such modest financial incentives are not nearly enough to overcome the disinclination to move of older, established workers with families, homes, employed spouses or children living at home, and roots in the community. To cover the full cost of a move for such a worker—including temporary living costs for a job search in a distant area, transportation, and losses on sale of a house—would certainly be extremely expensive. It seems likely that more socially productive uses of the resources could be found, for example, to create new employment opportunities in the areas where the people already live.

JOB SHARING

During periods of widespread economic distress, such as the 1981–1982 recession, I think it would be desirable to force some job sharing by cutting the official standard work week. The length of the standard work week is, of course, the point beyond which overtime pay rates begin. An employer would therefore have strong inducements to spread the available work among more workers. This should be done selectively, by the Secretary of Labor, for distressed industrial sectors (specified by Standard Industrial Classification). The determination of distress would be based on the current operating level of the industry as compared with the nominal capacity of that industry. Thus an industry might be defined as distressed when its operating rate drops below 70 percent of capacity. Below this point the standard work week could be cut back according to a schedule such as the following:

Industry Operating Rate (%)	Work Week (hours)
> 70	40
65–70	38
60–65	36
55–60	34
50–55	32

The scheme might be resisted by the most senior workers (who would stand to lose income), but it would keep union funds and unemployment insurance funds from being depleted as severely as at present. By the same token it would greatly reduce the sacrifice made by younger workers.

The principle might also be applied in the reverse direction to abate the inflationary impacts of cyclic periods of high demand. When the operating rate of an industry rises above 80 percent (for instance) it might make sense to *lengthen* the standard work week, up to, say, 42 hours, rising perhaps to 44 hours at 85 percent capacity.

JOB CREATION

It is perhaps fortunate that the need to create a fairly large number of jobs for unemployed men in their late thirties, forties, and fifties in the north central and northeastern part of the United States coincides with an urgent need to repair or replace thousands of canals, bridges, thousands of miles of water and sewer lines, and so on. The railroad rights-of-way, too, are badly deteriorated. Much of this infrastructure was originally built in the late nineteenth century and is now worn out. Allegheny County, surrounding Pittsburgh, has nearly 2,000 bridges, some of which are now unsafe and on the verge of collapse. Most of the other urbanized areas of the region have comparable problems.

The overall magnitude of the needed extraordinary reconstruction effort—above and beyond current levels of capital investment in highways, bridges—has been estimated as $900 billion on a national basis (Chapter 1). Expenditures on infrastructure currently amount to around $50 billion.

Suppose that we decided to increase the present outlay by 50 percent, or $25 billion more per year, during a catch-up period of perhaps a decade. This money could provide roughly 1 million additional jobs, both in construction and in supplier industries.[5] Adding to this the "multiplier effect" on local retail stores and other local services, an injection of additional spending of this magnitude could put 2 million or more people back to work, mainly in the region where relief is most needed.

Federal expenditures are not—and never have been—distributed in such a way as to return to each state the taxes that were collected there. But it is worth pointing out that year in and year out, on the average, more tax money is collected in the northeast and north central states than is returned to those states via federal expenditures. Based on Reagan's 1984 budget as submitted to Congress, per capita defense expenditures in the United States would be $539. But the regional distribution is very unequal: western states ($797), south ($722), northeast ($353), and midwest ($197). The effect is enhanced of course by the multiplier effect mentioned above. Today, however, the south and southwest are doing well—except for some localized economic problems they are booming—whereas the industrialized Great Lakes states suffer a continuing recession. A 10-year infrastructure catch-up program might help compensate for the inequitable distribution of defense expenditures.

Additional federal money must be allocated to the north central and northeast for reasons of equity:

1. Most of the nation's job losses are concentrated in this region, while most of the future gains in growing industries are likely to be elsewhere, unless these states can provide competitive attractions for new business investments.

2. It is impracticable and unfair to expect older, recently unemployed men who are heads of households, with mortgages to pay and children in school, to move away from their homes in search of jobs—especially when there are few jobs available anywhere. The situation is especially poignant in view of the fact that there are virtually no potential buyers for houses in dying cities like Youngstown, Fort Wayne, and Detroit.

3. The regional infrastructure is badly deteriorated and needs renovation, but the states cannot provide the funds in their present condition.

On the other hand, federal assistance in the form of grants for catch-up infrastructure renovation should be strictly contingent on an explicit social contract involving affected local unions. The terms of such a social contract should include the following provisions:

• Projects should be defined and laid out at the state level but under federal guidelines distinguishing catch-up projects from others.

• State and local governments should both contribute some of the costs to catch-up jobs.

• Contracts should be let and supervised by state-created, nonprofit regional agencies—not by the federal government. The federal role should be that of an auditor.

• Affected unions must agree to open shop (to permit unemployed workers from other unions equal opportunity to work) and "no strike" contracts, agreeing, in advance, to wages and working conditions for the duration of each catch-up job. Disputes to be arbitrated.

• Hiring rules for catch-up projects (only) to favor older unemployed workers living in the area of the project.

The point of all this is to provide both welfare benefits and public infrastructure improvements where they are needed, and to accomplish necessary tasks in the region, without permitting a particular group such as the construction unions leverage to capture unwarranted monopoly benefits for its own members. Without such a social contract, costs would rise and benefits (jobs) would not necessarily be made available to the individuals most in need of help.

The existing unemployment insurance (UI) system is a problem in itself. This system involves annual outlays in excess of $20 billion, depending on the unemployment rate. It is perversely generous to the footloose temporary or short-term worker who can earn unemployment coverage for a fixed period of time (e.g., 26 weeks) by working for as little as 6 weeks in a quarter. Yet the system offers nothing more than this to the worker with 10 to 20 years on the job, a mortgage, and a family to support[6] who is stranded by a plant closing. The system is obviously open to abuse, which makes it unnecessarily costly to employers,[7] yet unemployment insurance gives far too little protection to older workers with family responsibilities and many years of productive contribution to society.

EDUCATION, TRAINING, AND RETRAINING

The federal approach to job training since the Great Society program of President Johnson is largely irrelevant to present needs. Its primary emphasis was on providing entry-level skills (and work habits) to disadvantaged minorities. It was not designed to assist experienced workers to upgrade their skills or develop new ones. Indeed the training programs of the 1970s tended to concentrate on basic work habits and remedial education (reading, arithmetic, English for the foreign born). To the extent that skills were taught at all, they were the same low-level skills that are most likely to be displaced by robots and other forms of automation. This set of programs has been scaled down and partially defederalized by the Reagan administration, but there has been no significant redirection to date.

One expert states the situation as follows: "The employment and training needs of the dislocated workers are shared generally not with disadvantaged workers, but with the majority of the mainstream work force" (Bendick 1982: 33).

The point being made here is that the problems of dislocated workers, especially semi-skilled operatives, are not unique to that group, but are common to the majority of the work force. The problem, in brief, is inadequate general educational backgrounds and lack of marketable job skills on the part of most American workers, employed or not. If this point of view is accepted for the moment, the next questions are why workers are undereducated and what can be done to correct the situation. The answer to the first question is really a reflection on the quality of the public schools in this country. It would take us too far afield to comment further on the causes or possible cures of poor-quality public education.

Of course workers may recognize and rectify their educational inadequacies later. Or employers might do so. But most employers have very little incentive to provide generalized skills, since employees may very well take their upgraded skills elsewhere—or use them to bargain for higher wages. Notwithstanding this disincentive, employers actually spent $35 billion in 1981 on employee education and training. Clearly the socially optimal level would be considerably higher.

Workers also can and do choose to finance further training for themselves. But this may be impossible to afford if the worker is unemployed at the time. Even if the worker is employed, it requires a major cash investment as well as a commitment of spare time, to a project with an uncertain payoff. The uncertainty may be compounded by lack of reliable information about trends in the labor market to allow an intelligent choice of specialization. Nevertheless a large number of people do finance midcareer retraining. Again, however, it is clear that the amount being done is suboptimal.

The existing unemployment insurance system needs to be reconsidered. It should be thought of as a program to give workers positive incentives to accept new technology and to improve their own skills—whether they are actually unemployed or not. It should also be neutral to employers, eliminating existing disincentives to reemploy laid-off workers on a temporary basis.[8]

Several options within the unemployment insurance (UI) system have been proposed over the past decade or so:

- Allow the UI system to loan money to laid-off workers for retraining, the loan to be repaid by a surcharge on the individual's unemployment insurance contribution after reemployment.

- Tax-free employer and employee contributions to a fund to be used for retraining (or as a pension supplement after the worker retires).

- Unemployment insurance taxes could be abated for employers who outplace or retain workers as an alternative to layoff.

All of these schemes have some merit, but none has yet been subjected to thorough analysis with regard to eligibility, solvency, efficiency, and the new federalism—among others.

Options outside the unemployment insurance system include

- The Job Training Act (which replaced CETA) could be modified to include an explicit focus on retraining of displaced workers.

- Federal support (through the Department of Education) for adult vocational educational programs could be increased significantly—a good idea but totally inadequate in the scale so far proposed.

The complex structure of the federal government—both congressional committees and executive agencies—tends to favor piecemeal tinkering with existing programs, rather than new initiatives. Nevertheless the economic stakes are such that a single comprehensive program might well be the best answer. Such a program might be called the "Workers' Bill of Rights" (recalling the GI Bill of Rights introduced after World War II), the key features of such a plan include the following:

- All employed workers and their employers would contribute (tax free) to a fund replacing existing UI, to finance education and training.[9] Each worker would build equity in the system in proportion to the contribution made in his behalf. Credit would be transferable from job to job. The fund would pay tuition and fees at recognized vocational schools and colleges and provide supplementary income. A worker need not be unemployed to benefit.

- Accrued benefit rights in the event of displacement would be proportional to cumulative contributions.

- Short-service workers with little equity who are displaced involuntarily would be entitled to borrow from the fund with repayment deducted from later contributions.

- Ideally the program should be self-financing. However an initial endowment from general revenues should be not ruled out, notwithstanding current budgetary stringencies.

- Employers could elect to provide actual education/training services in house and deduct the cost from contributions to the fund (for those workers being trained). This would minimize or eliminate any net costs to employers with extensive training programs already.

- The problem of finding jobs for handicapped, disabled, unskilled, uneducated, and disadvantaged minorities who have been productively employed should be dealt with by different means. One possibility is a new version of the Civilian Conservation Corps (CCC) of the 1930s. Another is tax subsidies for firms willing to hire and train inexperienced workers. Reagan's Enterprise Zone concept may also have some promise. Existing welfare laws with their perverse disincentives to work in low-paying or part-time jobs should be replaced by a negative income tax. The minimum wage should also be repealed.

The Workers' Bill of Rights, by contrast, would emphasize mid-career retraining and educational upgrading. Two or three decades from now it should not be unusual for men and women in their thirties, forties, and fifties—with families—to be attending vocational schools, college, and graduate schools full-time for 1-year, 2-year, and even 3-year programs to earn formal degrees. One or more major career changes during a working life should be the rule rather than the exception.

The relevance of a program such as the foregoing—not to mention other proposals to create new support for high-technology education (e.g., Botkin et al. 1982) can be challenged. While today's fastest growing occupational categories, such as data processing-machine mechanics, computer systems analysts, computer operations, and office machine and cash register servicers, involve high-tech skills, the total number of new jobs in these areas will be small compared to other low-skilled job categories. In fact none of the top twenty occupations projected by the Bureau of Labor Statistics to generate the most jobs over the period 1978 to 1990 involves advanced technology.[10]

What of the oft-alleged coming shortages of electronic engineers and computer programmers (Botkin et al. 1982)? What about the fact that the USSR and Japan now educate many more mathematicians, scientists, and engineers than the United States? The implications are unclear. On the one hand it is clear that there is no shortage of scientists or engineers in the United States at present—if there were a real shortage, far fewer technically qualified people would give up their professions and switch to law or business school. It is also fairly clear that, while the USSR has a large pool of qualified engineers, most of them are used as technicians or bureaucrats. On the other hand Japan seems to find good jobs for its engineers, and the results speak for themselves. There is clearly a real shortage of high school math and science teachers in the United States but the explanation for this seems obvious and the cure easy. The shortage exists because public school teachers' unions, especially the National Education Association (NEA), refuse to allow any salary differential based on subject matter expertise. They will try to use the math-science teacher shortage as a lever to raise salaries uniformly for all teachers, regardless of qualifications, or teaching performance, and notwithstanding the fact that there is a surplus of qualified instructors in most subjects. The answer, of course, is for boards of education to insist on reasonable minimum educational qualifications for math-science teachers and to pay whatever salary differentials are necessary.

As regards the larger picture, reasonable people (like Levin and Rumberger 1982) can argue that currently available evidence suggests

- While there will be a significant percentage growth in high-tech jobs, the supply from normal sources can easily keep up with the projected demand.

- On the whole, job skill requirements are not being upgraded by the current wave of technological changes in factories and offices, but—if anything—downgraded.

The argument, in short, is that advanced automation may have a de-skilling effect on jobs. Though it is sometimes argued (e.g., Engleberger 1981) that machines or robots take over the most tedious, dangerous, and least skilled tasks first, leaving more challenging jobs to be done by human workers, there is reason to be skeptical. As was pointed out in connection with the product life-cycle model of industrial growth (Chapter 3), labor-skill requirements increase at first from general-purpose machine-specific in the very early stages of production to special-purpose product-specific during the rapid growth phase. But as the product is standardized, more automation is introduced, requiring less-skilled machine-tending labor. In effect the skills are incorporated in the machines. Case studies in specific industries such as printing confirm that newer forms of equipment do require less skill than the older forms. In offices the advent of word processing equipment seems to be reducing the level of skill required of typists.

Does all this mean that advanced education itself is becoming obsolescent? The question is worthy of serious consideration, though I answer it unhesitatingly in the negative. The basic reason for this answer is that education, in general, increases the adaptability of people to their environment and their ability to communicate and interact effectively with other people. It may well be true that many kinds of specialized manual crafts, such as welding or steam fitting, may not be needed so much in the future. It is also true that such skills have typically commanded higher pay than operating a cash register, assisting a nurse, or waiting on tables.

However, whereas a general education may seem irrelevant to a manual worker, it does make him more flexible and easier to retrain as specific skill requirements shift. Obviously education increases the effectiveness of persons involved in service occupations requiring interpersonal communication skills. A good restaurant waiter, for instance, exemplifies very high-level skills that are largely neglected in the United States, but require years of formal training in France. It also seems inevitable that pay scales for such jobs will be readjusted upward in due course, as slack demand for products of smokestack industries holds down wage increases in conventional blue collar occupations while rising demand for services gradually raises the competition and the level of compensation. Unionization may speed the process, of course.

The level of education required for most service sector jobs is somewhat arbitrary. The general rule is The more the better. Thus some jobs that call for a high school graduate today may well require a college graduate a few decades hence, because the customers (being college graduates themselves) demand it.

What about advanced education of scientists and engineers? The skeptical view is that any problem will be taken care of automatically by the marketplace. If a shortage develops, according to this argument, salaries will be bid up and more students will be attracted to the field. One counterargument, put forward by Botkin et al. (1982), is that high salaries from industry will attract the underpaid teachers and professors away from academia into industry, thus making it more difficult to quickly increase the supply of students.

This counterargument has some validity, at least in the short run, as secondary schools and universities can already attest. In the past few years they have had some difficulty, at times, attracting young mathematicians, chemical engineers, electrical engineers, and computer scientists as teaching faculty. The answer according to Botkin et al. is to start now to create a reserve supply of people with advanded training in certain critical fields.

Another, stronger argument exists for putting more emphasis on advanced technology in our society. It is based in effect on the idea that supply creates its own demand (Say's law in economics). In more mundane terms, if we educate more people in science and engineering — within limits of course — they will find ways of doing more science and engineering. For a start more scientists and engineers will work their way up the business and government hierarchy, and the effects of that (in my opinion) can only be beneficial. There will be more invention, more pressure to innovate and, in the end, more innovation.

UNION POLICY

It is unfortunate that both managers and workers tend to think of automation, especially robots, almost exclusively in terms of saving labor. That industry does so is revealed clearly by the fact that, despite many public assurances that no worker has lost his job to a robot, the financial justification for purchasing a robot nowadays is almost invariably and often exclusively based on labor hours saved. Workers and their unions, not being blind or stupid, soon get to know this. Blue-collar operatives, in particular, are beginning to feel a chill of apprehension about their future. Understandably, workers in many industries are beginning to seek a means of resisting the expected onslaught of "steel-collar" workers. The mechanisms being considered include tighter work rules, notification requirements, increased labor voice in shop floor decisions, inclusion of robot operators and programmers in the bargaining unit, shorter work weeks, more personal paid holidays, and so on.

Unfortunately all of these mechanisms will, to the extent they are adopted, further increase labor costs, further reduce productivity and exacerbate the long-run problem of U.S. noncompetitiveness. Worse, none of these measures will really be effective in protecting jobs and therefore in reducing workers' fears. In fact many of the restrictive mechanisms noted are already part of United Automobile Workers' (UAW) contracts with the major automobile manufacturers. Indeed the high costs of the protections in the UAW contract is partly responsible for the permanent closing of many plants and the loss of all jobs in these plants—regardless of length of workers' service to the firm. Even priority for job openings in other plants is small compensation to long-service employees, especially when seniority is not transferable to other bargaining units.

As a rule it is impossible for a worker to transfer seniority rights for promotion and for protection against job layoffs from one bargaining unit to another.[11] Seniority rights in these two critical areas are usually forfeited if the worker transfers out of the bargaining unit.[12] In some contracts seniority is specific to particular work areas within the plant. There are even cases where these rights are only retained if the worker remains with a specific occupation within the bargaining unit. Locals of the same union in different plants will not allow a worker to transfer seniority rights for promotion and for protection against layoffs. Needless to say, seniority is never retained if the worker joins another union.

Nontransferability of seniority is one of the most effective impediments to labor mobility, since it inhibits upgrading of skills, especially among older employees. It is a problem that only the labor unions themselves can solve.

WORKER OWNERSHIP

Another more controversial possibility that should be considered seriously by workers and their unions is to encourage their members to exchange wage and fringe benefits for equity in economically troubled industries. The Chrysler bail-out plan as originally proposed called for employees to receive up to 40 percent of Chrysler stock in exchange for wage and benefit give-backs. This proposal was rejected by the UAW, partly on the grounds that Chrysler stock at the time was extremely depressed, with a low market value, and partly because the union leaders saw a conflict of interest in being involved in management. (The workers do now own about 10 percent of the company's shares.)

In retrospect, however, it must appear to Chrysler workers—especially those who are now unemployed—that they missed out on a very good thing. Chrysler stock was the biggest gainer on the New York Stock Exchange for 1982—up almost 600 percent—because of the successful belt-tightening effort that was carried out under the direction of new chairman Lee Iacocca. The workers should have had a bigger share in this windfall, if only to compensate for some of their other sacrifices.

A somewhat similar, though more modest, agreement has recently been negotiated between the Pittsburgh-Wheeling Steel Company and its unionized workers. In exchange for giving up part of the existing wage and benefit package, the workers will receive stock in the company. Obviously there are some potential pitfalls, not least the relative unsophistication of many of the affected workers. Yet the agreement is promising, if only because it reduces the adversarial nature of the relationship between workers and management.

An even more radical approach has been taken by National Steel Company, as an alternative to closing its Weirton, West Virginia plant.[13] In effect the company offered to sell the entire plant (with 1982 sales of $817 million) to its 5,500 workers under an Employee Stock Ownership Plan (ESOP). For the workers the deal includes a 32 percent pay cut, of which 12 percent will be paid as stock in the new firm. Clearly success is far from guaranteed, but the chances seem reasonably good because virtually everyone in the town of Weirton is strongly committed to the survival of the mill as a viable enterprise. For National Steel, the deal is better than simply closing the plant would have been.

Within a decade many other older steel mills will likely either be closed or operated by workers who have reluctantly become owners. But for the latter option to be economically viable, good faith negotiations must be initiated before a terminal crisis occurs. United Steel Workers (USW) will have to encourage and facilitate these negotiatins, or they will not occur.

The workers of Amtrak, the government-owned railroad system built on the ruins of the Penn-Central and several other northeastern lines, are in the process of bidding for ownership of the refurbished system. They will probably succeed.

In other mature basic industries too, this option is worth exploring. General Motors has already spun off one small subsidiary (making bearings) to an employee group,[14] and the plant is apparently a modest success under worker-ownership. There is no basic reason why some other auto parts plants faced with potential closure should not be offered a similar option. There is nothing magic about worker ownership, but it may be preferable to unemployment. However, both experience and logic suggest that worker owners are likely to be more motivated to modify work rules, work longer hours and to take pay cuts, if necessary, to assure a firm's survival, than hired hands would be. Sometimes these factors can make the difference between survival and failure.

If the idea became popular, of course, it might result in demands for ownership by workers in economically healthier parts of the economy. Some conservatives see this possibility as a threat to the capitalist system itself. Until now, top-level union leaders have shared this negative view. It is now time for both unions and management to reconsider. A number of worker-owned companies already do exist and many of them have been very successful. In a firm where workers have little or no ownership interest, there is no mechanism for ordinary employees to become aware of the real financial condition of various operations, or the options open to management. Under these circumstances union strategy in the past has essentially been to try to protect its membership by negotiating contractual constraints on management options. This sort of protection is illusory.

The result of several decades of such negotiations in the basic industries, most union contracts are a bureaucratic nightmares of complex work rules and grievance procedures. Managers have relatively little scope to manage or innovate at the plant level, which is one of the chief reasons labor costs often become uncontrollable. Higher level decisionmakers may then be faced with starker choices of the "go/no go" variety. The day finally comes, usually during the low point of an economic cycle, where the only feasible option is "no go": eliminate a shift or close a plant. This often occurs because all less extreme options have been ruled out by the terms of labor agreements, and it simply is not possible to explain and revise the complex contract in a short time. The extreme difficulty of renegotiating contracts in an adversarial climate is amply illustrated by recent experiences in the steel industry and in the automobile industry.

The real problem in the present union-management relationship can be traced to inadequate information and institutionalized lack of trust. Unionized workers and their traditional leaders typically regard themselves as being fundamentally *in opposition* to management. For this reason, workers tend to be highly skeptical of information emanating from management sources. Management, in turn, tends to be unnecessarily secretive about many matters of concern to workers, for no obvious reason except traditional suspicion. In these circumstances it is virtually impossible for unions and management to reach satisfactory agreements.

On the other hand if workers own—directly, or through their pension funds—a sufficient share of the stock of a firm to elect one or more directors, the information gap largely disappears. Directors are legally entitled to *all* relevant information about the activities, policies, and financial condition of a firm.

Board members have a fiduciary responsibility to select operational managers and provide policy guidelines to them, in the interest of the shareholders. It is consequently ethically and even legally problematic for a board of directors to justify actions in the general public interest or in the interest of employees if those actions cost money with no clear evidence of benefit to the owners (stockholders) of the corporation. Clearly a board may be forced to take actions to ensure the long-term survival of the firm, at some cost to current employees. This applies equally to a hypothetical union representative on the board as to a banker, lawyer, or retired executive. This is clearly an uncomfortable position for the union representative. Nevertheless, in that position he or she would be able to provide far more meaningful help to his membership than sitting across from an adversary at the bargaining table but lacking necessary knowledge of the real options.

NOTES TO CHAPTER 8

1. Not, I should like to add, at the expense of the rest of the world. The world economy is not a zero-sum game where gains to one must be losses to others. An economically healthy and growing United States will be a much better market for the exports of other countries than will a depressed and stagnant United States.

2. Not necessarily different people. Much of the wealth of the United States is owned by pension funds, money funds, banks, insurance companies, and other institutions on behalf of savers—namely all of us. We may receive income as wages or as dividends or interest.

3. Examples where both union and nonunion jobs exist in parallel are easy to find: for example, construction workers, bus and truck drivers, teachers, assembly-line workers. Union construction workers, when employed, typically get about twice the wage of nonunion workers doing the same job.

4. The right to strike may be an important aspect of our cherished civil liberties—even if it is costly in economic terms and leads to inequitable distribution of income.

5. Quite a lot of the jobs would be in producing structural steel, cement, and machinery. But these industries are largely located in the north central region, where unemployment is currently the worst.

6. Except the nominal weekly allowance (about $5) per dependent.

7. The costs are passed on indirectly to employers, whose unemployment insurance rates are raised as claims are made against the fund. For an experienced employer with a history of 3 years or more, rates can rise as high as 30 percent of payroll costs—a crippling burden for almost any business, particularly a small one. This fact also provides a strong disincentive to call back laid-off workers. Employers often prefer to pay overtime to their remaining employees. The existing system, operating as it does through the states also results in states with high unemployment having the highest unemployment insurance rates—a major disincentive to employment-creation in these states.

8. Disincentives exist in the present system, because under present rules temporary reemployment entitles the employee to additional benefits that in turn increase to the employer's unemployment insurance rates.

9. A similar scheme has been operational in France since 1971. Under the French Further Vocational Training System, employers are obligated to spend 1.6 percent of payroll either in-house, as contributions to external educational institutions, or as a payroll tax to the government for retraining purposes. (See Bendick 1982.)

10. In fact the five biggest growth categories are janitors, nurses' aides, sales clerks, cashiers, waiters and waitresses. These job categories account for 13 percent of the total projected employment growth between 1978 and 1990. Only four of the "top 20" categories require education beyond a high school degree and only two require a college degree (teaching and nursing). (Carey 1981).

11. Seniority rights for other privileges, such as vacation preferences, health care, pensions, or for overtime, are more easily transferred across bargaining units within the same company.

12. If a union such as the UAW has national agreements with a large company, there are exceptional circumstances under which seniority rights might be transferred. When this does happen, the worker is often greeted with hostility by other workers in the plant to which he transfers.

13. The Weirton plant, though fairly up-to-date, became unprofitable because (1) its chief product, tinplate, has a declining market, (2) its nonunion workers were paid 10 percent more than the United Steel Workers' pay scale, and (3) EPA demanded a major investment in pollution clean-up.

14. Again the alternative was to close the plant within 3 years, due to the switch from rear-wheel to front-wheel drives (which require a different type of bearing).

9 ENERGY AND MATERIALS POLICY

In an ideal world of peaceful free trade and competitive markets, we might not need an explicit policy on energy and materials, but we do not live in that world. In pursuit of their perceived national interests, most governments have interfered with the play of market forces in significant ways and continue to do so. The U.S. government is no exception. One big problem in the United States is that de facto energy and resource policy is the accidental consequence of policies that were mostly put in place long ago for other purposes.

A brief listing of some of the key elements of current U.S. energy and materials policy would have to include the following:

- Oil and mineral depreciation allowances to encourage rapid exploitation of raw materials in the ground.

- Very low taxes on gasoline (compared to Europe and Japan).

- Price controls on interstate natural gas and, for a decade until 1981, on domestic petroleum.

- Regulation of rates charged by electric power utilities, based on concept of fair rate of return on invested capital.

- Vast public road-building projects, but little and last-resort federal support for railroads.

- A far-flung network of military bases and armed flotillas around the world to protect U.S. allies and interests.

223

At first sight some of these items may not seem to be related to energy and materials scarcity or availability or price. But they all are. The depreciation allowance obviously favors producers, but in a highly competitive market lower costs to them are passed on to consumers. The consumer has been the major beneficiary of depreciation allowances and of price controls on gas and oil. However, the other side of the coin is that extensive consumption is encouraged. The United States consumes two to three times more energy per capita than other industrialized countries, and U.S. manufacturers are less energy-efficient than foreign counterparts and competitors, thus—indirectly—accelerating actual depletion of high-quality resources. In addition to this failing, the depletion allowance provides no incentive at all to develop technological alternatives to resources that are beginning to run out. It is simply left to the private sector and the federal government to invest in long-range research on such things as breeder reactors, fusion, solar energy, and so on. There is at present no economic mechanism to link the intensity of the federal funded research effort to the urgency of the problem. Long-range private R&D in these areas scarcely exists.

Another consequence of price controls, especially on natural gas, was an apparent shortage. During the 1950s and 1960s most gas was produced as a by-product of petroleum drilling for the simple reason that the controlled price of gas was too low to justify exploring or drilling for gas per se. Hence U.S. gas discoveries began to dwindle when U.S. petroleum discoveries dwindled, and a shortage was proclaimed. For several years in the late 1960s, gas distribution systems were not permitted to add new residential or industrial customers.

An unintended and unfortunate consequence of the way in which electric utilities in the United States are regulated is that, paradoxically, they have had very little incentive to conserve energy. In fact any cost savings they happened to achieve would automatically be passed on to consumers by the regulatory agencies (public utility commissions), while any proposed expenditures to increase efficiency could not be recouped by higher rates, except in unusual circumstances.

On the other hand electric utilities have had strong incentives, in the past, to do whatever they could to *increase demand* for electricity. In this case the utility is required by law to meet the demand, and the public utilities commissions must allow a fair rate of return on the new investment. Note that the larger the investment base, the greater the allowed total profit from captive ratepayers. Spurred on by this perverse incentive system, electric utilities actively promoted electric appliances (even acting as retail outlets) and when the gas distributor stopped taking new customers, electric utilities actively promoted electric heat for homes as a "clean, convenient" alternative to oil heat.

Moreover, to some extent confusing cause with effect, the electric utilities convinced themselves that demand for electricity would continue to grow indefinitely, as it had grown since 1950, at a rate of 7 percent per annum. At this growth rate, capacity must be doubled every 10 years, which implied an eight-fold increase in electricity consumption from 1970 to 2000. The utilities accordingly began to order large numbers of nuclear power plants. Since 1974, as the 7 percent growth rate dropped to 3 percent or less, many of the planned nuclear plants have been canceled.

The policy of building and maintaining highways with public funds while *not* subsidizing railroads has had three major consequences. One is that the United States now has a very efficient truck-based goods distribution system. The system is fast, flexible, and cost-effective—if highways are regarded as public goods. The second consequence is that cities have spread out and factories have moved away from their old center city locations (where they were served by docks or railyards) and have relocated near the major freeways, to facilitate movement by trucks. By spreading out employment, a low-density housing pattern has evolved, permitting people to exchange living space for distance. Low-cost land and cheap building materials encouraged large, single-family dwellings. These structures are, incidentally, poorly insulated and require a great deal of energy to maintain. The third consequence is the virtual demise of the once efficient passenger rail transport system, except in the Washington–Boston corridor. Trolleys and commuter rail lines have practically disappeared except in New York, Philadelphia, and Chicago. People overwhelming drive cars to work.

The net result of all these policies—none of which had future energy availability as its focus—is that the United States now consumes far too much energy, especially gasoline. We also consume too much electricity, relative to other fuels, and we consume too much natural gas for very inefficient space-heating, because the price of gas was set much too low for several decades. Another indirect consequence is that the United States—having an artificially inflated appetite for energy—has now become dependent on foreign sources of supply, at least for petroleum. This fact has undoubtedly had an impact on our concept of national interest and the cost of defending it against potential threats.

AVOIDING OBSESSION WITH LIFELINES

It is clear that in the short to medium run, it is very important for the United States to have access to foreign sources of critical materials, including petroleum as well as cobalt, manganese, chrome, bauxite, tungsten, and the platinum group metals. The problem is to make quite sure we count *all* the costs of securing access to the foreign sources in question. A significant share of the national defense budget is really allocated to this function. The costs of maintaining a major military presence in the Indian Ocean—a navy task group, airfields and three naval bases (Oman, Somalia, and Diego Garcia), a disastrous former alliance with Iran, and several uneasy military-political marriages of convenience (e.g., South Korea, the Philippines, Egypt)—as well as the Rapid Deployment Force based in the United States—should be counted as indirect costs of maintaining access to Persian Gulf petroleum supplies. *This is in fact an indirect but massive government subsidy to users of imported petroleum.* If alternative sources such as synfuels could be obtained more economically, they should be exploited instead.

One of the dangers to be avoided, if possible, is confusing political alliances with reliable trading partners. It is often presumed, without much analysis, that a shift of temporary political allegiance by some Third World country from the United States to the USSR would automatically result in a cut-off of trade. There is absolutely no evidence to support this notion. On the contrary, the Soviets themselves have been perfectly willing—indeed eager—to sell natural gas, chrome, manganese, and platinum group metals to the West, on long-term contracts. Soviet-backed Angola and Iraq are as happy to sell oil to the West as mavericks like Libya and Iran—and are more reliable about it. When Marxist guerillas based in Angola overran Zaire's cobalt-producing Shaba province in 1977-78, the price of cobalt soared in anticipation of a cutoff. But the production was unaffected and cobalt became a glut. The only (briefly) successful embargo of recent years was the Arab oil embargo of winter 1973-74, which was organized by our "friends," the Saudis! Lately it has been the United States that is reluctant to trade with countries in the Soviet orbit. The United States has also proven itself to be a less reliable trading partner than the Soviet Union, a fact that is beginning to cost us dearly in export sales.

More to the point, perhaps, embargoes do not work, whether our own or the other side's. The United Nations' sponsored embargo of Rhodesian chrome was an embarrassing failure, as was the U.S. wheat embargo against the Soviets. The realities of international trade are such that buyers with hard currency, especially dollars, can always obtain supplies of commodities they need. The price of cobalt or chrome may double or triple for a few months or even a couple of years, but for any commodity whatever, except petroleum, the total financial cost to the U.S. economy of a period of artificially high prices is likely to be much less than the cost of a single nuclear-powered aircraft carrier. It must be borne in mind, too, that even a very big jump in the price of cobalt, tungsten, or chromium could not really affect the final price of finished jet engines or machine tools very much.

The most effective policy instrument to reduce the economic threat of embargoes, cartels, or production interruptions due to other causes (such as local wars) is probably stockpiling. Fortunately the United States already has an extensive mineral stockpiling program. Petroleum is also now being stockpiled in the strategic petroleum reserve. The stockpile typically contains from a few months' to several years' requirements of critical materials. Historical experience suggests that stockpiles of this magnitude are more than enough to cushion the impact of almost any embargo or foreign production interruption.

For longer term energy and materials problems, the best answer is to develop technological substitutes. In many cases there are feasible substitutes already on the shelf and available on short notice whenever the price of the traditional material rises enough to justify the shift. Later prices typically drop sharply as the original material re-enters the market. For instance, when supplies of tungsten from China were cut off from the United States during the Korean War, machine tool builders quickly substituted alloys of molybdenum. A year later high-speed tungsten steel was largely replaced as a tool material by tungsten carbide requiring tungsten in much smaller amounts. The price of tungsten dropped rapidly and permanently after a couple of years of artificially high tungsten prices. As a result the U.S. stockpile's 4- to 5-year supply of tungsten is probably grossly excessive.

An even more dramatic example of rapid substitution occurred during World War II, when the Japanese conquered Indo-China and Malaya, the world's major source of natural rubber. The fear that this might happen had played a significant role in U.S. policy toward Japan, but in the actual event factories to produce synthetic rubber were built very rapidly in the United States and natural rubber was largely displaced. Again the substitution was permanent. Similarly, the Germans succeeded in synthesizing ammonia in 1914 in response to the British control over supplies of Chilean nitrates. In World War II the German chemical industry successfully synthesized aviation gasoline from coal via the Fischer-Tropsch process, to compensate for lack of petroleum. The likely long-term substitute for gasoline in the United States is methanol, from natural gas or coal.

In summary, the best long-term protection against resource shortages is a strong technological capability for developing viable substitutes. From a policy perspective the problem is to stockpile technological capabilities: theoretical knowledge, facilities, and engineers.

RESOURCE-RELATED RESEARCH NEEDS

The long-term resource problem is one that is inherently in the domain of the federal government. Very high priority should be given to a long-term research program aimed at processing lunar materials and manufacturing in space, with the ultimate objective of planting a self-sustaining industrial activity on the moon or in geosynchronous orbit. This should be done independently of any military considerations. A government-sponsored consortium could be an appropriate vehicle for this activity.

In the same spirit, the U.S. government should support a much bigger long-range research program aimed at recovering resources from the oceans. This program should include intensive exploration, the development of mobile submersibles, processes to recover minerals from the sea bottom, and processes to extract materials from ocean water. A consortium would be appropriate for much of this, too.

The need for long-term substitutes for petroleum and other fossil fuels has not been eliminated by the recent and very temporary oil glut. The urgency of building full-scale synfuels plants is a bit less than was thought a few years ago because of the unexpected effectiveness of conservation, but the effort should not be dropped from the R&D agenda. Nor should research on breeder reactors be stopped. Both of these programs have suffered from an attempt to freeze designs prematurely and to build large plants. They should continue at a moderate level, nevertheless.

However, much higher priority should be given to fusion research and direct conversion of solar energy to electricity via photovoltaic cells. The problems of manufacturing such cells might be addressed by a research consortium of electronics firms with government assistance. The present level of effort is too low and progress is too slow.

How is all this research, development, and demonstration to be financed? The best answer may well be to earmark funds from a small tax on both extraction and imports of all raw materials, including energy.[1]

Free market die-hards tend to assert dogmatically that the market will "automatically" take care of the future scarcity problem via higher prices, which justify riskier long-term investments. But the theory is not well-founded in this case. It is true that growing scarcity will eventually drive the price of petroleum up, but the response of the economy to higher prices is not knowable, except qualitatively. There are deep technological uncertainties the market can not really evaluate (nor can the technologists themselves). Nobody—certainly not Wall Street—really knows, for example how long it will take to develop methanol-based propulsion systems as a viable alternative to petroleum as a motor fuel, or how much investment will be required.

The following set of measures would ensure that R&D effort is linked to tangible evidence of resource scarcity:

1. Reduce the existing resource depletion allowances progressively.

2. Impose a small severance tax on extraction of virgin domestic resources (say 3 percent on nonenergy materials and 10 percent on fuels).

3. Impose a somewhat larger (say 6 percent) duty[2] on imported nonenergy raw materials and on the *resource content* of imported processed materials. This would rise to 20 percent for hydrocarbons.

4. Create an independent but congressionally chartered Technology Trust Fund to receive these revenues, to be administered by a

board of trustees similar to that of a foundation. The members might be appointed for long staggered terms by the President, but subject to approval by the National Academy of Sciences, as well as the Senate.

5. The Technology Trust Fund should support three basic missions:

    a. Support of education at all levels, with emphasis on improving science/engineering training.

    b. Support basic science at Universities and applied R&D through major research institutions such as the NASA laboratories and via competitive funding of specific projects.

    c. Venture capital investments for new high-technology firms, profits from such investments to be recycled in a revolving fund.

In addition to promoting the foregoing purposes, the proposed tax would indirectly serve three additional special purposes, namely:

    d. Discouraging raw materials and energy consumption, hence slowing down the rate of actual resource depletion.

    e. Increasing incentives to recycle existing materials. Waste and pollution would thereby be reduced at the source.

    f. The Trust Fund would support all federal R&D programs that are now paid for out of general tax revenues. It would, ceteris paribus, permit these tax revenues to be used for other purposes, or to cut the federal deficit.

Many people will argue against this proposal on the grounds that it is politically infeasible. Perhaps this will ultimately prove to be the case. But a similar scheme has been implemented successfuly in Japan, and even the Reagan administration has proposed a per barrel tax on imported petroleum. Indeed the citizens and voters might be wiser than their political representatives give them credit for being. The problem is urgent, because a number of fledgling enterprises developing innovative alternative sources of energy are currently in danger of being put out of business as a result of the so-called petroleum glut, even though the glut will disappear very quickly if any kind of sustained economic recovery takes place. Then we would be even more vulnerable to petroleum blackmail than before, since private developers of alternative forms of energy, having been once burned, are likely to be very slow to return to the arena.

NOTES TO CHAPTER 9

1. The tax in question has been proposed numerous times in the past—notably by President Carter—and rejected by Congress. However, the linkage suggested here between an energy resource tax and energy/resource R&D has not been proposed before. In Japan energy research is funded by taxes on gasoline, electricity, and other fuels. The underlying justification for this linkage is similar to the theoretical argument for a mineral depletion allowance—namely that as a resource is exhausted the capital needed for replacement should be created automatically. Depletion allowances have undoubtedly generated funds for exploration and drilling, as was intended.

    Unfortunately depletion allowances—especially for petroleum—have also reduced real costs of production and indirectly constituted an effective subsidy for consumers. Low-priced petroleum in the United States has stimulated excessive consumption by large automobiles, poorly insulated houses, and so on.

2. The computation of direct and indirect resource content is complicated but feasible.

10 THE INNOVATIVE CORPORATION

CORPORATE STRATEGY

One clear implication of the characterization of the international competitive situation in Chapters 1 and 2 is that it now makes little sense for firms located in the United States to try to compete against foreign firms, either in foreign markets or in the U.S. market, on the basis of product standardization and cost minimization by exploiting hard automation and economies of scale to reduce costs. This strategy has had its day and can no longer be successful in the long run for products where the technology of production is itself standardized and is generally available from multinational capital goods producers.

Yet many firms, perhaps dazzled by the spurious simplicity of the concept of the experience curve, according to which costs fall as a function of cumulative production, have pursued a strategy of price-based competition through *market-share maximization*. The tactical means employed in support of this strategy include accelerated product standardization and premature commitment to mass production. This strategy has been adopted most enthusiastically by the U.S. semiconductor industry in its zeal to compete with Japanese producers.

That the new mass-production facilities tend to be located in off-shore export processing zones (EPZs) is regarded as a regrettable fact of life by many corporate managements. There are obviously operational disadvantages to the system, including political uncertainty, currency fluctuations, sheer distance, and language and cross-cultural interfaces within the management structure. In many cases Japanese manufacturers are able to produce at lower cost than American firms precisely because they do not have these problems.

There are two further strategic pitfalls for U.S. firms. First, over-hasty commitment to product standardization and mass production can easily backfire if the technology is still rapidly evolving. Ford's Model T was an example. Second, this approach puts U.S. based multinational corporations increasingly in conflict with U.S. workers, U.S. communities, U.S. taxpayers and the long-run U.S. national interest. Such a schizoid situation cannot last indefinitely. The growing conflict of interest between stockholders of U.S. based multinational corporations and the U.S. work force and general public will inevitably in a democracy give rise to a demand for corrective legislation. It is only too likely that, given the plethora of special-interest groups that seems to confuse and complicate the legislative process, the end product could turn out to be ill-advised and counter-productive. Organized labor is likely to seek higher tariffs, tougher antidumping rules, capital export taxes or controls, mandatory severance pay for displaced workers, and more. The essential point is that the basic strategy now being followed by most U.S. corporations is inherently prone to failure.

In contrast to the prevailing market-share maximizing price-based strategy, which might be termed *accelerated maturity*, the appropriate approach for U.S. based firms could be dubbed *extended product adolescence.* The key elements of this alternative strategy would include

- Increased emphasis on new product development (to delay standardization).

- Increased emphasis on *flexible*, rather than hard, automated production facilities, with maximum integration of design, prototype, and manufacturing processes, through CAD, robotics, and CAM.

- Competition on the basis of product *performance* and quality, rather than product *price.*

It cannot escape the notice of anyone who follows the fortunes of major corporations that firms that constantly develop and introduce new products perform better over the years than firms that have followed the standardization and mass-production route. Many of the latter are now troubled or defunct.

Of course many U.S. firms, foreseeing product obsolescence, have tried to compensate by diversification through acquisition. Results vary widely, but many conglomerates today appear to be in serious difficulty, not only because of clashes between different corporate cultures but because a firm with several obsolescent products is probably even worse off than a firm with only one.

What if the firm has existing cash cows producing a standard commodity-like product by a standard single-purpose technology? As long as the marginal cost of production allows some profit, such a business can obviously continue to operate. But the production facility itself will seldom be worth replacing, at least in the United States.[1] Every business has a finite life expectancy. It is a primary responsibility of top management to assure that mature and aging products are eventually replaced by viable successors. Unfortunately, many managers have neglected this aspect of their jobs because of the prevalence of short-term financial measures of performance. Managers are discouraged by the prevailing financial incentive system from making long-run investments (in new product development and manpower retraining) at the expense of short-run profits.

The problem of upgrading and retraining the workers is a separate issue, which most firms in the United States regard as a responsibility of the public sector. Although the federal government cannot ignore the problem and must take a major responsibility for solving it, the usual corporate view has been myopic. The most successful large firms in the future are likely to also be the most progressive in a social sense, seeing their experienced workers as long-term assets of the firm, not to be discarded lightly.

CORPORATE R&D POLICY AND MANAGEMENT OF TECHNOLOGICAL CHANGE

Corporate goals, along with profitability and growth, include *survival*, although this goal is seldom spelled out explicitly. The function of R&D in a modern corporation is to provide a flow of technology in support of all its goals. Thus long-term survival and growth require the continuous introduction of new products to replace obsolescent ones.[2] Profitability, on the other hand, may be improved either by increasing demand, especially where there is excess capacity, or by reducing costs. R&D contributes to the first via product improvement—to be distinguished from new product development—and to the second by process improvement.

Product or process technology may be acquired from two basic sources: *internal*, from within the R&D laboratories, or *external*. In the past most major U.S. firms have attempted to develop almost all their own product technology internally, although process technology is quite commonly purchased from capital goods producers. In a few cases this closed-door policy was perhaps justified on the grounds that the United States in general and the firm in particular already led the world technologically and therefore had little to gain from the outside. Such a claim of autonomy can no longer plausibly be made by very many U.S. firms, if any.

It makes a limited amount of sense for a firm with a technological advantage over all its competitors, domestic and foreign, to screen out most offerings from the outside world on the grounds that (1) no firm can take up all options or investigate everything and (2) the firm is already exploring the most promising avenues. On the other hand it does *not* make sense for a firm that is technologically "one of the pack" to behave in this way. Nevertheless traditional corporate R&D departments continue to raise effective barriers against accepting ideas from outside, rather than reaching out for the best ideas, regardless of their source. Often this is justified on the basis of a grossly exaggerated idea of the firm's technological prowess vis-à-vis its competitors.

It is reasonable to assume that the traditional behavior pattern is reinforced by the existing R&D resource allocation system.[3] In-house scientists and inventors see external sources of ideas as competing for funds needed by their pet projects. Exhortation alone is unlikely to alter this pattern. The most promising approach is to modify the structure and funding of the R&D function as a whole.

The traditional structure typically includes four major activities:

1. Basic research (optional)[4]
2. Applied research focused on new products and processes
3. Product engineering (closely associated with design and marketing)
4. Process engineering (closely associated with manufacturing)

A more effective structure would include an additional explicit function, namely technology acquisition and adaptation (TAA). Staff could initially be taken from existing activities. The TAA department would be in business strictly to search the world for new technologies applicable to the firms' products and processes. It would continuously monitor U.S. and foreign publications and patents. It would regularly send representatives to conferences, trade shows (especially abroad), and—where practicable—to visit R&D laboratories of universities, government agencies, suppliers, competitors, and customers.

In short the TAA department would formalize the "gatekeeper" function that has been identified as playing an important (but largely unrecognized) role in technologically innovative organizations. It would also retain responsibility for adapting and developing the ideas and technologies from outside sources to fit the needs of the firm. TAA would thus compete with the traditional applied research department, with its emphasis on creating new technology uniquely for the firm. The performance of each of the two departments would thus be measurable over time in terms of its cost-effectiveness as a provider of new technology for the firm.

The integration of product engineering and production engineering also becomes extremely urgent if the adolescence-extending strategy advocated earlier is adopted and the pace of technological change accelerates. Flexibility in this sense requires more than programmable multipurpose machines and production processes. It also requires formal integration of the design engineering and manufacturing functions within the firm. Computer-aided design (CAD) generates a great deal of useful digital information. Ideally this information would be used as a basis for automatically developing control software for the manufacturing process. But the typical manufacturing process, especially for metal products, is a complex sequence of operations and at the present time the only way of arriving at the most efficient sequence of operations is by trial and error, guided by much hands-on experience. Many of the decisions that must be made in the course of optimizing a manufacturing process are not even quantifiable at present. Thus it will be many years, perhaps two decades, before the processes of design and manufacturing can be linked by computers permitting two-way communication of information.

But the difficulties make it all the more urgent that these functions be integrated organizationally. This will be a very traumatic change for many U.S. firms, where product design and engineering at present are prestigious jobs closely associated with the corporate research laboratories and typically carried out in isolated parklike physical settings—while manufacturing engineering is a function with less prestige carried out by low-status engineers stationed in cubicles near the plant floor working with foremen and technicians. Interaction between these groups is made more difficult by physical distance and by very different perceptions of design priorities. Yet more effective interaction between design and manufacturing *must* be achieved in any U.S. based corporation that is to survive the 1980s as a technological leader. Indeed the problem is deeper. Management in the 1980s and 1990s will be, increasingly, *management of technological change*. Technological considerations must be involved in every decision, at every level of the corporation, as intimately as financial considerations are today. The business world in the United States is utterly unprepared for this change. Indeed many large U.S. firms have no technological experts among their top management groups. This is a deficiency that will have to be remedied. The time available for management restructuring is very short.

THE HIDDEN DRAG OF ACCOUNTING PRACTICES

The next major issue is the tendency of most U.S. firms to maximize short-run profits at the expense of long-run growth or even long-run survival. See, for example, Hayes and Abernathy 1980. Based on a survey of practices in a number of large firms, Cyert and DeGroot (1982) assert that just six variables are currently used as a basis for decisionmaking by top management. These six are (1) net current earnings per share, (2) net current dollar sales, (3) current cash flow, (4) current return on investment, (5) current return on stockholder's equity, and (6) new orders received. It is significant that all are strictly financial and short-term in nature despite the fact that maximizing short-term financial results can very easily lead to disastrous consequences. As one example, it can be argued that Braniff Airline's demise resulted from a high-risk strategy of maximizing current profits by expanding its service and route structure in a period of high demand and high interest rates (1978–79). It did this by increasing its debt equity ratio from 65 percent in 1978–79 to 97 percent by April 1982 when it filed for bankruptcy. (Interestingly this common measure of risk is *not* on the "big six" list.

The tendency to focus on the very short run has been widely blamed on the rapid spread through industry of cut-and-dried project evaluation methodologies, using financially oriented accounting systems based on oversimplified models of the firm, commonly taught in graduate schools of business and practiced by the ubiquitous MBA's. The former dean of Carnegie-Mellon University's Graduate School of Industrial Administration has commented on this issue

> Despite the growing complexity of analytic procedures developed for cost accounting applications, the particular manufacturing setting used as the basis for virtually all modeling efforts remained remarkably simple, even naive. The production model involved the manufacture of a set of products with known input requirements of labor, materials, and capital. The products are mature so that elaborate standards have usually been developed for the usage and cost of all input factors. Analytic procedures were used to produce summary financial measures of performance to understand the first order effects of price changes, usage variations, and volume effects. Profit planning was enhanced by first separating costs in fixed and variable components and then rationalizing the product mix given available capacity and current market conditions. Sophisticated inventory models were introduced and procedures developed to estimate relevant cost parameters so that "optimal" production quantities and inventory levels could be computed. Uncertainty was formally incorporated by assuming a probability distribution for sales quantities, or price and cost parameters, and determining the consequent probability distribution for profit. Various summary measures of uncertainty such as the expected value of perfect or sample information and expected opportunity losses could be computed. Statistically quality control models were adapted to aid the variance investigation decision. The decision and control relevance of alternative cost allocation procedures for service departments and for joint product situations were also developed during this time.
>
> Thus, by the early 1970's, an impressive array of analytic procedures had been mobilized to enhance the management accounting practices in an extremely simple and stable manufacturing setting. (Kaplan 1982)

However, while the international competitive failure of many U.S. firms is demonstrable, its cause is still debatable. The core problem may be traceable to the competitive system of compensation and promotion that is widely practiced in the United States and the management account system that is used to measure performance. With a few exceptions, most firms try to identify future top management candidates early and deliberately move them around from job to job to give them a wide experience. The fast-track managers seldom stay in a given job more than 2 or 3 years. To stay on the fast track, new managers must demonstrate significant improvement in profitability during their brief tenure. To underline this point, managers often get significant bonuses—which they come to depend on—based on current performance in relation to predetermined targets. If there is a setback, even a temporary one, the manager may very well lose the bonus, his or her place on the fast track, or even the job.[5] In effect, the line manager has a personal discount rate much higher than the official corporate rate used for financial evaluations of capital investments. To borrow Kaplan's words once again:

> I believe that an important contributing factor to the relative deterioration in U.S. manufacturing productivity arises from an excessive preoccupation by U.S. executives with short term financial performance and a lack of attention to continually upgrading manufacturing performance through improved production procedures and the introduction of new technology. In recent years, bonus plans for senior executives based on financial performance (either accounting or stock-price based measures) have become prevalent. Many bonus plans are a function of accounting income each year, and even those with a "long-run" outlook (e.g. performance shares), are based on earnings growth over a three-to-five year period. U.S. executives have even succeeded in devising compensation packages that provide rewards almost unrelated to the ability to earn superior returns on shareholders' equity. With many executives believing that their stock price is a function of their most recent quarterly earnings report, executives with stock options, stock appreciation rights and phantom stock plans are similarly concerned with the effects of their decisions on reported earnings. (Kaplan 1982)

The incentive and control systems of most firms at divisional levels support the concern of senior management with short-term financial performance. Budgets and performance measures are typically denominated in financial, not physical, measures. These measures are compatible with the background and training of almost all accountants and controllers—specialists who probably have never been exposed to nonfinancial performance measures. Thus U.S. firms, increasingly led by senior executives with finance or legal backgrounds, focus on short-term financial performance, while their counterparts overseas are modernizing their productive facilities, providing better training to their labor force, and increasing the quality, reliability, and deliverability of their products.

The system as practiced in most U.S.-based firms is counterproductive, for a very fundamental reason. The reason is that the average line manager is extremely reluctant to sacrifice, or even jeopardize, current results—regardless of possible future benefits—since the benefits are likely to be credited to his successor. This strongly inhibits the use of new production technologies, even proven ones, because it is impossible to introduct a new technique or an unfamiliar new machine, such as a robot, onto the production line without temporarily halting production at least briefly. It has been found in case after case that such interruptions always cause some retrogression in the learning or experience curve. Retrogression, of course, means temporarily reduced profits or even losses—anathema to most business-school-trained managers.

The fact is that some temporary retrogression inevitably accompanies technological progress, because progress by definition involves a change in the status quo. If managers are unwilling to risk short-term costs or losses in order to achieve long-term improvements, the best interest of the firm will suffer.

As Kaplan pointed out, management accounting systems almost universally used today are appropriate for monitoring the current status of a *mature* business, where both products and production technology are standard and competition is on the basis of minimizing costs (Richardson and Gordon 1980). Indeed the usual accounting framework is appropriate *only* for this situation. It is notably unsuited to measuring the effectiveness of executive or corporate performance in new product development, product improvement, and new market penetration. It is unsuited to measuring the flexibility of a manufacturing system or the responsiveness of a firm to changes in the environment.

The performance of executives and divisions of firms with new or adolescent products is typically measured by the same criteria that are applied to established and standardized products. The consequences can be extremely counterproductive: Product designs are prematurely frozen and standardized when, in fact, the long-term corporate strategy should be the reverse, namely to promote continued technological change and extended adolescence. The inappropriateness of applying traditional accounting measures to technological matters also leads to other systematic management errors. One of these is the "sailing ship phenomenon"—a tendency on the part of management to wait too long to replace an established technology by a new one. A great many former market leaders have failed to respond to technological opportunities (or threats) and suffered the consequences. More generally, traditional accounting methods lead to underinvestment in long-range research, because most of the value of the new knowledge acquired is indirect—it contributes in unexpected ways—which makes it, by definition, unaccountable.

How can top management motivate middle management to lengthen planning horizons for the good of the firm as a whole? For that matter, how can boards of directors motivate top management in this direction? Possibilities include

- Productivity-based bonus incentive plans for production workers, e.g., Scanlon plans (Moore and Ross 1978)

- More explicit emphasis on promotion from within (to discourage career job-hopping)

- Incentive plans for management tied to strategic goals, not financial goals, using performance measures appropriate to the stage of the product life cycle (Rappaport 1982)

- Longer term appointments (5 years) for a particular management slot

- Tenure (up to age 55 or so) for employees with 5 to 10 years experience, as in Japan

- Deferred bonuses based on longer term results. (A manager might receive a "standard" salary, based on length of service, and additional deferred compensation based on the average performance of his or her unit for a period starting perhaps in the third year of tenure and terminating 5 years after the manager moves on or retires.[6] This would give each manager a greater incentive to help his successor get off to a fast start.)

All of the foregoing schemes would reduce the pressure on managers to product instant results. This would occur if compensation and promotion policies took into account longer term results of managerial decisions.

To optimize such a complex scheme might introduce significant practical difficulties of measurement and allocation of credit for innovations. However, even an imperfect scheme along these lines would probably be better than the present system. Of course a subtler and less financially oriented management accounting system would be of enormous benefit.

If the present system of executive compensation were effective (even in maximizing short-run results) there should be a correlation between top management rewards and corporate performance. In other words the executives of the most successful corporations should be getting the most money. In fact there appears to be almost no relationship between executive pay and company growth of profitability. According to *Fortune* (July 1982), any such relationship is largely coincidental. Corporate size is a much better predictor of chief executive pay than profits—which is one reason why so many chief executives concentrate their efforts or negotiating favorable mergers or acquisitions, rather than on building growth from within. Incidentally, chief executives' pay has escalated much faster in the last decade than the wages or salaries of ordinary workers, while corporations themselves have performed worse. This fact is not hidden from lower paid workers and has unquestionably contributed negatively to employee morale in many firms.

It is difficult to escape the conclusion that something is wrong at the center: in the board room. What seems to be wrong is that most boards of directors are effectively chosen by the operating management, reporting to and controlled by an all-powerful chairman. There is little or no effective external control over executive performance. Unhappy stockholders in practice have only one option—to sell. Even unhappy board members have little control over events, as was clearly evidenced in the 1982 four-way takeover battle between Bendix, Martin-Marietta, United Technologies Corporation, and Allied Corporation. Board members of Bendix were asked to approve complex multibillion dollar proposals with as little as 2 hours notice, (*Wall Street Journal*, November 24, 1982).

A good solution, recommended by Peter Drucker (1974) and others, would be the creation, by legal mandate if necessary, of truly *independent* boards of directors for large public companies. Such boards probably should represent (at least) the various constituencies large corporations must deal with, including suppliers, customers, banks, local governments, and labor—as well as stockholders. What is at issue is the difference between democracy and totalitarianism in the business world. Drucker's plan would indeed be a revolutionary change, but perhaps a necessary one if other changes discussed in this book are to be successfully implemented.

DECENTRALIZATION

Solutions to productivity problems in U.S. business lie mostly in the hands of government and top management, in roughly equal proportions. This is not to downgrade the importance of incentives all the way down the hierarchy. By all means let industry experiment with Scanlon plans, quality circles (QC's), and other devices. The only hard fact in the firmament is that the present system, as it operates in most manufacturing firms, is rapidly becoming untenable. Employees at the lower levels generally feel that they are being treated as pawns by unsympathetic foremen, unresponsive middle managers, and manipulative top managers whose motives are primarily personal aggrandizement rather than the general welfare. Even Lee Iacocca is not much of a hero to the rank and file at Chrysler.

What can be done? In the long run the inflexible hierarchical bureaucratic structure itself will have to give way. The large centralized bureaucracy is the end result of a historical trend that simply went too far. Communications and feedback through many layers of management is just as inefficient in industry as it is in government (where businessmen see the problem quite clearly). Decentralization is the wave of the future, I believe. The long-resisted breakup of AT&T is likely to result in a set of parts more valuable separately than they were as a whole, because of deregulation and elimination of some internal subsidies. What is good for AT&T, in this case, might well be good for many other large firms, especially conglomerates that were constructed in the 1960s and 1970s out of diverse and incompatible components.

It is not necessary to go all the way to devolution (as in the AT&T case) to gain some of the benefits of decentralization. The model currently used in many conglomerates is for the corporate headquarters to behave toward its subsidiaries like an investment banker. Teledyne Corporation is a very good example. Typically the corporate headquarters staff allocates capital to its subsidiaries and takes a share of their revenues, on the basis of purely financial considerations, such as profit and return-on investment targets. But all operating decisions are left to the subsidiaries as long as they meet financial targets set by the corporate staff.

This type of organization is probably a step in the right direction, but I think the time will come when the role of headquarters will be still further reduced, and the employees will have to be given more responsibility in the running of the plant or business unit where they work. The problem with management by setting short-term financial targets has already been discussed. Ultimately some form of at least partial employee ownership or other means of benefit sharing will be found to provide the best incentives and therefore the most efficient production.

This will evolve naturally, as other successful social innovations do, by imitation of successful examples. Already a number of companies are largely owned by their own pension funds, or by employees, or both. AT&T is one such. The number will probably grow. The rapid spread of employee stock ownership and purchase plans (ESOPs) is a clear indicator of the direction of evolutionary development of the U.S. economic system.

NOTES TO CHAPTER 10

1. There are some exceptions to this rule, notably producers of building materials, glass, paper, and other bulky products where low domestic raw material costs preclude foreign competition.
2. The product life-cycle may be measured in months (as in womens' fashions) or in centuries (steel rails), but the usual range is between one and three decades.
3. The usual policy is to reject any idea "not invented here."
4. This is probably more important in a leading-edge laboratory (such as Bell Laboratories). Many firms leave basic research to government and universities, while still claiming nonexistent R&D superiority.
5. At the same time, there is a good deal of "gaming" with choice of targets. The targets top management teams select for themselves—which obviously govern targets for lower echelon managers—are often on the low side, presumably to make the bonus predictable and easy to earn. As a result middle managers often go so far as to stockpile surefire improvements against a rainy day (a threat to the bonus) rather than introducing them immediately.
6. A line manager under this scheme might temporarily share credit for the performance of his unit with his predecessor. I am happy to note that Alan Greenspan recently publicly proposed this kind of scheme, at least for chief executives. As this book is going to press I have learned that General Motors is planning to modify its bonus system to incorporate some of these features.

11 PULLING THE POLICY STRANDS TOGETHER

Policy put into practice has complex side effects. It is all too easy to recommend raising motor fuel taxes sharply and to forget the impact on taxi drivers and truckers—especially the independents. Deregulating natural gas may result in some elderly pensioners freezing to death in a cold winter. Improving income tax reporting and collection of tips may cause a regulatory burden for small restaurant proprietors. Catching welfare cheats by toughening eligibility rules can also result in serious hardships for the handicapped or for the working poor. Easing regulatory constraints on business may preserve jobs for some but result in toxic waste dumping in the backyards of others. And so it goes. The fact is that somebody gets hurt almost regardless of what government does—or doesn't—do.

It is sometimes easier for leaders to do nothing than to risk the certain ire of those who will be adversely affected by an active program. As Olson (1982) points out, the beneficiaries are seldom comparably grateful, especially if the benefits are spread widely. But it is clear that a policy of minimum government is no longer a feasible option for our nation. The real options are much narrower.

The common theme in all of the policy recommendations in this book is to encourage technological innovation, From our rural forebears we inherited the carrot and stick metaphor for encouraging desired behavior in a recalcitrant donkey. This simple choice may work where the desired behavior is very well-defined, but innovation

247

is a more complex activity. The carrot is the psychological or financial reward for success. The stick is the pressure of economic (or geopolitical) competition. Government can improve the flavor of the carrot by providing R&D subsidies, incentives to venture capital, or by helping inventors to defend themselves against infringers. Government can also help force innovation—say in motor vehicles—by imposing emission standards, fuel consumption standards, or thermal insulation standards, to mention just three of many possibilities. More important, it can refrain from protecting consumers by subsidizing the use of scarce materials (such as petroleum) and thus discouraging the development of alternatives.

Nonetheless the primary area for policy intervention at the present time is neither carrot nor stick. It is—retaining the original metaphor—to fill the potholes and smooth the donkey's path. Thanks to a number of established societal rewards for success—carrots—the donkey wants to move forward, all right, but he's stuck in the mud. Much of the mud was unintentionally created by government itself (not always at the federal level). The task before us is therefore to pave the muddy road.

The mud that prevents technological progress is a host of institutionalized impediments to change. Inflexible mass-production technologies based on "dedicated" special-purpose machines, inflexible union work rules, and inflexible government regulations (such as building codes that specify exactly the materials and construction techniques that are permitted and thus virtually rule out innovative materials and methods) inhibit or prevent progress. Extraordinarily high long-term debt/equity ratios and interest rates constitute a parallel constraint on financial flexibility.

Alleviating the rigidity of inflexible mass-production technology means changing the practice of using mechanical controls for high-output special-purpose machines so costly and specialized that it is impossible to alter product design once the machines are in place. The advent of programmable general-purpose automation promises eventually to improve the economics of large-scale manufacturing. This in turn would release the constraint on technological innovation by large-scale manufacturers. The competitive benefits to U.S. industry seem fully to justify a federal investment in R&D on flexible automation.

In the same vein, the resource and energy proposals in this book are entirely focused on reducing our national dependence on so-called critical imports, especially where the continued availability of such a resource is being subsidized or ensured by military power. Bearing in mind that the British economic decline can be attributed, in part at least, to the burden of maintaining a far-flung system of resource supply from its Empire, whereas the Germans were motivated to seek technological subsitutes, U.S. policymakers will be far better off to eschew the dangerous temptation to protect existing sources in unstable parts of the world and to choose instead, to subsidize long-range R&D on alternatives.

The majority of the human resource policy proposals here are rooted not in humanitarianism but in the recognition that most of the existing rigidity in the labor market is the result of fear of technological obsolescence and the natural desire of workers to organize themselves in such a way as to protect their jobs. Unfortunately, much of the hard-won protection is illusory; worse it has caused inefficiencies that became part of the problem. To persuade workers to give up their dependence on narrowly defined job classifications and work rules, they must be offered something better: a commitment from society as a whole that they will *not* be victimized and discarded by the processes of technological change. This necessary reassurance is not provided effectively by the existing unemployment insurance system, which is also unduly burdensome on small and growing businesses and in areas most in need of attracting new enterprises to replace failing ones. The necessary commitment necessarily involves a variety of kinds of assistance—including relocation, retraining, and upgrading displaced workers' education. It also may involve increased worker participation—even ownership—in existing mature businesses.

The industrial policy and technology transfer proposals are part of this broad commitment. We as a society cannot reassure our fellow citizen-workers that they will not be victimized unless we create the right incentives for corporate employers. This is a complex and difficult area in which it is easy to go wrong by protecting a few at the expense of many. The first priority is to avoid the kind of phony protection for workers in declining industries that simply shifts all the burden to the rest of society. This is a formula for accelerating our economic decline, rather than reversing it.

What we as a society can do is to imitate the successful Japanese Ministry of International Trade and Industry by creating a government agency to help U.S. firms compete internationally. The means include sponsoring trading cartels for handling certain imports and exports, antitrust exemptions for multifirm R&D projects, and protecting infant products from predatory competition. As a quid pro quo for temporary infant-product protection, there should be constraints on the export of new technology via licenses or turnkey plants. This notion will be anathema to advocates of free trade, but it is really only a slight modification of the accepted notion that inventors should have a period of freedom from competition as a societal reward for their risky investment of time and money. The point is that a nation grants such rewards only because it is conceived to be in the wider interest of the nation—all citizens, including workers and taxpayers—to do so.

Finally, the role of large corporations was addressed in this book. The problem is again rigidity. Abernathy (1978) characterized it aptly as "the productivity dilemma." Having been very successful by exploiting economies of scale and mass production, U.S. firms have evolved into specialized organizations for this purpose. Unfortunately the specialized mechanisms that worked very well in this context-including extreme division of labor and hierarchical control based on purely financial measures of performance—are peculiarly unsuitable for technologically innovative smaller firms.

A large organization originally oriented to mass production may not be able to restructure itself successfully into a dynamic organization oriented to technological innovation. However some of the requisites of an innovative organization can be identified. They include (1) a strategic plan emphasizing technological change, (2) an open door to technology developed outside the firm, (3) effective integration of product development, engineering and manufacturing functions, and—on top of everything else—(4) management incentives based on long-term results. Such a radical restructuring of the firm must begin in the boardroom, and to encourage change some government intervention might well be justified to bring about more effective outside representation on boards of directors. Legally mandating the inclusion of representatives of labor and public interest groups may be effective. Ultimately most of the large firms of today should and will, of their own accord, break up into smaller, more homogeneous, more manageable units with substantial worker ownership. The law facilitating employee stock ownership and purchase (ESOP) plans is a step in the right direction. Government policies to speed up this decentralization should probably be given serious consideration.

REFERENCES

Abernathy, William J. 1978. *The Productivity Dilemma.* Baltimore: John Hopkins University Press.

Abernathy, William J.; Kim B. Clark; and Alan M. Kantrow. 1981. "The New Industrial Competition," *Harvard Business Review* (September–October).

Abernathy, William J., and James M. Utterback. 1975. "A Dynamic Model of Process and Product Innovation." *Omega* 3, no. 6: 639.

Abramovitz, Moses. 1956. "Resources and Output Trends in the United States since 1870." *American Economic Review* 46 (May).

Abramovitz, Moses, and Paul A. David. 1973. "Reinterpreting Economic Growth: Parables and Realities." *American Economic Review* 63 no. 2 (May).

Agnew, William. 1980. "Automotive Fuel Economy Improvement." General Motors Research Laboratories Pub. GMR–3493, November.

Alexander, A. J., and J. R. Nelson. 1973. "Measuring Technological Change: Aircraft Engine Turbines." *Technological Forecasting and Social Change* 5: 189–203.

American Machinist. 1978. *The 12th American Machinist Inventory of Metalworking Equipment, 1976–78.*

Arrow, Kenneth. 1962. "The Economic Implications of Learning by Doing." *Review of Economic Studies* 29 (June).

Ashton, T. S. 1948. *The Industrial Revolution, 1760–1830.* London: Oxford University Press.

Ayres, Robert U. 1977. "Public Goods, Efficiency and Environmental Statistics." In *Ecomometric Contributions to Public Policy*, Proceedings of International Economic Association Conference. New York: St. Martins Press.

Ayres, Robert U. 1978. *Resources, Environment, and Economics.* New York: John Wiley and Sons, Chapter 3.

Ayres, Robert U., and Steven Miller. 1982. *Robotics: Applications and Social Implications.* Cambridge, Mass.: Ballinger Publishing Company.

Baranson, Jack, and Harold Malmgren. 1981. *Technology and Trade Policy: Issues and an Agenda for Action.* Washington, D.C.

Beckman, Martin J., and Ryuto Sato. 1969. "Aggregate Production Functions and Types of Technical Progress: A Statistical Approach." *American Economic Review* (March).

Bell, Daniel. 1967. "The Year 2000: Trajectory of an Idea." In *Toward the Year 2000: Work in Progress.* Report of the Commission on the Year 2000. *Daedelus* (Summer).

Bell, Daniel. 1973. *The Coming of Post-Industrial Society.* New York: Basic Books.

Bendick, Marc, Jr. 1982. "Workers Dislocation by Economic Change: Toward New Institutions for Midcareer Worker Transformation." Washington, D.C.: Urban Institute.

Binswanger, Hans P., et al. 1978. *Induced Innovation: Technology Institutions and Development.* Baltimore: Johns Hopkins University Press.

Boretsky, Michael. 1966. "Comparative Progress in Technology, Productivity and Economic Efficiency: USSR vs. USA." Studies prepared for the Subcommittee on Foreign Economic Policy of the Joint Economic Committee, Congress of the United States, Washington, D.C.

Boretsky, Michael. 1973. "U.S. Technology: Trends and Policy." U.S. Department of Commerce.

Boretsky, Michael. 1977. "Technology, Technology Transfers and national Security." Delivered at National War College, February 26.

Boston Consulting Group. 1970. "Experience Curves as a Planning Tool." *IEEE Spectrum* (June).

Botkin, James; Dan Dimancescu; and Ray Stata. 1982. *Global Stakes: The Future of High Technology in America.* Cambridge, Mass.: Ballinger Publishing Company.

Brecher, Richard A. 1982. "Optimal Policy in the Presence of Licensed Technology from Abroad." *Journal of Political Economy* 90, no. 5: 1070–1078.

Brecher, Richard A., and Ehsan U. Choudhri. 1982. "The Leontiev Paradox, Continued." *Journal of Political Economy* 90, no. 5: 820–823.

Bryant, Lynwood. 1967. "The Beginnings of the Internal Combustion Engine." In M. Kransberg and C. Pursell, eds., *Technology in Western Civilization.* New York: Oxford University Press.

Business International Corporation. *The Effects of US Corporate Foreign Investment* 1960–1972 (2nd in series) and 1970–1980 (10th in series), New York.

Buzbee, B. L.; R. H. Ewald; and W. J. Worlton. 1982. "Japanese Supercomputer Technology." *Science* 218 (December 17).

Cameron, Rondo. 1967. "Imperialism and Technology." In M. Kranzberg and C. Pursell, eds., *Technology in Western Civilization.* London: Oxford.

Carey, Max. 1981. "Occupational Employment Growth through 1990." *Labor Review* 104 pp. 42-55. Washington, D.C.: Bureau of Labor Statistics.

Chandler, Alfred D. 1977. *Visible Hand: The Managerial Revolution in American Business.* Cambridge, Mass.: Harvard University Press.

Christensen, Paul P. 1981. "Land Abundance and Cheap Horsepower in the Mechanization of the Antebellum United States Economy." *Exporations in Economic History* 18: 309-329.

Churchill, Winston. 1958. *A History of the English Speaking People. Vol. IV, The Great Democracies.* New York: Dodd, Mead.

Cipolla, Carlo, and Derek Birdsall. 1980. *The Technology of Man.* New York: Holt, Rinehart and Winston.

Cochran, E. B. 1968. *Planning Production Cost Using the Improvement Curve.* San Francisco: Chandler.

Cooper, Richard N. 1971. "Technology and U.S. Trade: A Historical Review." In *Technology and International Trade.* National Academy of Engineering, Washington, D.C.

Cross, Ralph E. 1980. "The Future of Automotive Manufacturing—Evolution or Revolution?" *Automotive Engineering*, 75th Anniversary Issue (July).

Cunningham, James A. 1980. "Using the Learning Curve as a Management Tool." *IEEE Spectrum* (June).

Cyert, Richard, and Morris DeGroot. 1982. "The Maximization Process under Uncertainty." Working paper. Carnegie-Mellon University.

David, Paul A., and Th. van de Klundert. 1965. "Biased Efficiency Growth and Capital-Labor Substitution in the U.S. 1899-1960. *American Economic Review* 55.

Davis, John P. 1961. *Corporations.* New York: Capricorn Books.

Deane, Phyllis. 1979. *The First Industrial Revolution.* Cambridge, England: Cambridge University Press.

Denison, Edward F. 1962. "The Sources of Economic Growth in the United States and the Alternatives before Us." Supplementary Paper no. 13, Committee for Economic Development, New York.

Denison, Edward F. 1967. *Why Growth Rates Differ.* Washington, D.C.: Brookings Institution.

Denison, Edward F. 1979. *Accounting for Slower Economic Growth.* Washington, D.C.: Brookings Institution.

Diebold, John T., et al. 1951. "Making the Automatic Factory a Reality." Boston, Mass.: Harvard Business School.

Drucker, Peter. 1979. *Management: Task, Responsibilities, Practices.* New York: Harper & Row.

Eads, George C. 1983. Statement before the Joint Economic Committee, U.S. Congress, June 29.

Eads, George C.; Arthur Levitt Jr.; Thomas K. McCraw; Robert B. Reich; and Lester C. Thurow. 1983. In Alan M. Kantrow, ed., "The Political Realities of Industrial Policy." *Harvard Business Review* (September-October): 76-86.

Ellul, Jacques. 1967. *The Technological Society* (translated from French). New York: Vintage Books.

EPRI (Electric Power Research Institute). 1982. "Metals from Flyash." *EPRI Journal* (May).

EPRI (Electric Power Research Institute). 1982. "Forecasting the Patterns of Demand." *EPRI Journal* (December).

Fabricant, Solomon. 1954. "Economic Progress and Economic Change." *34th Annual Report of the NBER*. New York: National Bureau of Economic Research.

Fearnsides, John J. 1982. Foreword in Kannan, Rebibo, and Ellis. *Downsizing Detroit*. New York: Praeger.

Ferguson, Eugene S. 1967. "The Steam Engine before 1830." In M. Kranzberg and C. Pursell, eds., *Technology in Western Civilization*. New York: Oxford University Press.

Fieldhouse, D. K. 1961. "Imperialism: A Historiographical Revision." *Economic History Review* 14 (December): 187–209.

Ford, David, and Chris Ryan. 1981. "Taking Technology to Market." *Harvard Business Review* (March–April).

Fraumeni, Barbara M., and Dale W. Jorgenson. 1980. "The Role of Capital in U.S. Economic Growth." In G. von Furstenburg, ed., *Capital, Efficiency and Growth*. Cambridge, Mass.: Ballinger Publishing Company.

Freeman, Christopher. 1982. *The Economics of Industrial Innovation* 2nd ed. Cambridge, Mass.: MIT Press.

Freeman, Christopher; John Clarke; and Luc Soete. 1982. *Unemployment and Technological Innovation*. London: Frances Pinter.

GAO. 1976. "Manufacturing Technology—A Changing Challenge to Improved Productivity." Report of the Comptroller-General to the Congress, Washington, D.C., June.

GAO. 1982. "U.S. Military Co-Production Programs Assist Japan in Developing Its Civil Aircraft Industry." Report of the Comptroller-General to the U.S. Congress, Washington, D.C., March 18.

Galbraith, John Kenneth. 1958. *The Affluent Society*. Boston: Houghton Mifflin.

Galbraith, John Kenneth. 1978. *The New Industrial State*, 3rd ed. rev. New York: Mentor Books.

GM. 1982. *General Motors Public Interest Report*.

George, Peter. 1982. *The Emergence of Industrial America*. Albany: State University of New York Press.

Gilfillan, S. Colum. 1935. *The Sociology of Invention*. Chicago: Follet Publishing Company.

Gilfillan, S. Colum. 1937. "The Prediction of Inventions." In W. Ogburn et al. *Technological Trends and National Policy*.

Gordon, Theodore J., and T.R. Munson. 1980. "Research into Technology Output Measures." *The Futures Group* (November).

Graham, Edward. 1978. "Technological Innovation and the Dynamics of the U.S. Competitive Advantage in International Trade." In C.T. Hill and J.M. Utterback, eds., *Technological Innovation for a Dynamic Economy.* New York: Pergamon Press.

Gregory, Gene. 1982. "New Materials: Japan's High Technology Building Blocks." *Materials and Society* 6, no. 4.

Griliches, Zvi. 1964. *American Economic Review* 54 (December): 961.

Griliches, Zvi. 1971. "Price Indexes and Quality Change." *Studies in New Methods of Measurement.* Cambridge, Mass.: Harvard University Press.

Habakkuk, H.J. 1962. *American and British Technology in the Nineteenth Century.* Cambridge, England: Cambridge University Park.

Hayes, Robert. 1982. Quoted in *Technology Review* (May/June).

Hayes, Robert. 1982. "The Undermining of Business Credibility." *New York Times,* October 10.

Hayes, Robert, and William Abernathy. 1980. "Managing Ourselves into an Economic Decline." *Harvard Business Review* (August).

Hicks, John R. 1932, revised 1963. *The Theory of Wages.* London: MacMillan.

Hufbauer, Gerald C. 1970. "The Impact of National Characteristics and Technology on the Commodity Composition of Trade in Manufactured Goods." In Raymond Vernon, ed., *The Technology Factor in International Trade,* New York: Columbia University Press.

Hulten, Charles R., and James W. Robertson. 1982. "Corporate Tax Policy and Economic Growth: An Analysis of the 1981 and 1982 Tax Acts." Washington, D.C.: Urban Institute. Draft.

Illinois Institute of Technology Research Institute. 1968. *Technology in Retrospect and Critical Events in Science* (TRACES), prepared for National Science Foundation Chicago.

Institute for Energy Analysis. 1976. "The U.S. Energy and Economic Growth: 1975–2010." Oak Ridge Associated Universities.

Jacobson, Catherine. 1982. "The Technology Transfer Issue." *Business America* (May).

Jarrett, Noel; S.K. Das; and W.E. Haupin. 1980. "Extraction of Oxygen and Metals from Lunar Ores." *Space Solar Power Review,* vol. 1. SUNSAT Energy Council, pp. 281–287.

Kaplan, Robert S. 1982. "Management Accounting and Manufacturing: The Lost Connection," draft manuscript submitted for publication.

Kemp, Murray C. 1969. *The Pure Theory of International Trade and Investment.* Englewood Cliffs, N.J.: Prentice-Hall.

Kendrick, John W. 1956. "Productivity Trends: Capital and Labor." *Review of Economics and Statistics* (August).

Kendrick, John W. 1961. *Productivity Trends in the United States*, National Bureau of Economic Research. Princeton, N.J.: Princeton University Press.

Kendrick, John W. 1980. "Productivity Trends in the United States." In Maital and Metz, eds., *Lagging Productivity Growth: Causes and Remedies*. Cambridge, Mass.: Ballinger Publishing Company.

Kendrick, John W. 1977. *Understanding Productivity*. Baltimore: Johns Hopkins University Press.

Kendrick, John W., and Elliot Grossman. 1980. *Productivity in the United States: Trends and Cycles*. Baltimore: Johns Hopkins University Press.

Kondratiev, Nicolai. 1935. "The Long Waves in Economic Life." *Review of Economics and Statistics* (November).

Kravis, Irving; Alan Heston; and Robert Summers. 1981. "New Insights into the Structure of the World Economy." *Review of Income and Wealth* (December): 348–349.

Krugman, Paul. 1979. "A Model of Innovations, Technology Transfer and the World Distribution of Income." *Journal of Political Economy* 87, no. 2: 253–266.

Kuznets, Simon. 1953. *Economic Change*. New York: W. W. Norton.

Leamer, Edward E. 1980. "The Leontiev Paradox, Reconsidered." *Journal of Political Economy* 88 no. 3: 495–503.

Leonard, William N. 1971. "Research and Development in Industrial Growth." *Journal of Political Economy* (March–April).

Leontiev, Wassily. 1954. "Domestic Production and Foreign Trade: The American Capital Position Reexamined." *Econ. Internz.* 7 (February): 9–38. Reprinted in Caves and Johnson, eds., *Readings in International Economics*. Homewood, Ill.: Irwin.

Levitt, Theodore. 1965. "Exploit the Product Life Cycle." *Harvard Business Review* (November–December).

Louis, A. M. 1982. "The Bottom Line on Ten Big Mergers." *Fortune* (May 3).

Lynn, Leonard H. 1982. *How Japan Innovates: A Comparison with the U.S. in the Case of Oxygen Steelmaking*. Boulder, Colo.: Westview.

Maddison, Angus. 1969. *Economic Growth in Japan and the USSR*. London: George Allen and Unwin.

Magaziner, Ira C., and Robert B. Reich. 1982. *Minding America's Business: The Decline and Rise of the American Economy*. New York: Harcourt, Brace, Jovanovich.

Magee, Stephen P. 1977. "Multinational Corporations and International Technology Trade." In National Science Foundation *Preliminary Papers for a Colloquium on the Relationship between R&D and Returns from Technological Innovation*, May 21.

Magee, Stephen P. 1977. "Multinational Corporations and the Industry Technology Cycle and Development." *Journal of World Trade Law*.

Magee, Stephen D. 1977. "Technology and the Appropriability Theory of the Multinational Corporation." In J. Bhagwati, ed., *The New International Economic Order: The North-South Debate.* Cambridge, Mass.: MIT Press.

Mansfield, Edwin. 1968. *The Economics of Technology Change.* New York: W. W. Norton.

Mansfield, Edwin. 1972. "Contribution of R&D to Economic Growth in the United States." *Science* 175 (February 4).

Mansfield, Edwin. 1977. "Research and Development, Productivity Change, and Public Policy." In National Science Foundation *Colloquium on the Relationships between R&D and Economic Growth/Productivity,* November 9, Washington, D.C.

Mansfield, Edwin; A. Romeo; M. Schwartz; D. Teece; S. Wagner; and P. Brach. *Technology Transfer, Productivity and Economic Policy.* New York: W. W. Norton.

McCarthy, Michael D. 1965. "Embodied and Disembodied Technical Progress Functions in the Constant Elasticity of Substitution Production Function. *Review of Economics and Statistics* 47 (February).

Melman, Seymour. 1965. *Our Depleted Society.* New York: Holt, Rinehart and Winston.

Melmed, Arthur. 1982. "Information Technology for U.S. Schools." *Phi Delta Kappan* (January).

Mensch, Gerhard. 1979. *Stalemate in Technology.* Cambridge, Mass.: Ballinger Publishing Company.

Miller, Charles. 1971. *The Lunatic Express.* London: MacMillan.

Minasian, J. 1969. *American Economic Review* 59 (May): 80.

Moore, Brian, and Timothy L. Ross. 1978. *The Scanlon Way to Improved Productivity: A Practical Guide.*

Morris, James. 1968. *Pax Britannica: The Climax of An Empire.* New York: Harcourt, Brace and World.

Morrison, Elting. 1966. "Almost the Greatest Invention." In *Men, Machines and Modern Times.* Cambridge, Mass.: MIT Press.

Muller, Ronald E. 1980. *Revitalizing America: Politics for Prosperity.* New York: Simon and Schuster.

Multhauf, Robert P. 1967. "Industrial Chemistry in the 19th Century." In M. Kranzberg and C. Pursell, eds., *Technology in Western Civilization.* New York: Oxford University Press.

Mumford, Lewis. 1967. *The Myth of the Machine: The Pentagon of Power.* New York: Harcourt, Brace, Jovanovich.

National Academy of Sciences, Materials Advisory Board. 1966. "Report in the Principles of Research Engineering Interaction." Washington, D.C.

National Research Council. 1978. "Technology, Trade and the U.S. Economy," report of a Workshop at Woods Hole, Mass., August 22-31. Washington, D.C.: National Academy of Sciences.

National Science Foundation. 1971. *Research and Development and Economic Growth/Productivity Papers and Proceedings of a Colloquium*, NSF 72–303, December.

National Science Foundation. 1977. *Relationships between R&D and Economic Growth/Productivity.* Preliminary Papers for Colloquim, May 21.

Neary, Michael, and Thomas A. Wilson. 1982. "Chemical Process," U.S. Patent 4,355,160, Filed December 4, 1980. Issued June 15.

Nelson, Richard. 1962. Introduction to National Bureau of Economic Research, *The Rate and Direction of Inventive Activity.* Princeton, New Jersey: Princeton University Press.

Nelson, Richard. 1964. "Aggregate Production Functions as Medium-Range Growth Projection." *American Economic Review* 54 (September).

Noble, David F. 1977. "The Wedding of Science to the Useful Arts." In *America by Design.* New York: Oxford University Press.

Norsworthy, J. R. 1982. "Recent Productivity Trends in the U.S. and Japan." Testimony prepared for the Senate Subcommittee on Employment and Productivity of the Committee on Human Resources, Washington, D.C.: April 2.

Novak, Michael. 1982. "Why Latin America is Poor." *Atlantic Monthly* (March).

Noyce, Robert. 1977. "Large Scale Integration: What Is Yet to Come?" *Science* 195 (March 18).

NTIS. 1982. "International Development in Computer Science." U.S. Department of Commerce, National Technical Information Service, PB 82–233677, Washington, D.C.: June 1.

OECD. 1981. *Economic Outlook* (December).

Oettinger, Anthony. 1980. "Information Resources: Knowledge and Power in the 21st Century." *Science* 209 (July 4): 191–198.

Ogburn, William. 1937. Chairman of the Subcommittee on Technology, Report to the National Resources Committee, National Research council. *Technological Trends and National Policy, Including the Social Implications of New Inventions.* Washington, D.C.: National Academy of Sciences.

Ogburn, William, and Dorothy Thomas. 1922. "Are Inventions Inevitable? A Note on Social Evolution." *Political Science Quarterly* 37 (March): 83–98.

Olson, Mancur. 1971. *The Logic of Collective Action* 2nd ed., Cambridge, Mass.: Harvard University Press.

Olson, Mancur. 1982. *The Rise and Decline of Nations.* New Haven, Conn.: Yale University Press.

O'Neill, Gerard. 1981. *2081: A Hopeful View of the Human Future.* New York: Simon and Schuster.

OTA. 1982. *Genetic Technology: A New Frontier.* Washington, D.C.: Office of Technology Assessment.

Ouchi, William. 1981. *Theory Z: How American Business Can Meet the Japanese Challenge.* Reading, Mass.: Addison-Wesley.

Papert, Seymour. 1980. *Mindstorms: Children, Computers, and Powerful Ideas.* New York: Basic Books.

Polli, Rolando, and Victor Cook. 1969. "Validity of the Product Life Cycle." *Journal of Business* 42, no. 4 (October): 385.

Porat, Marc. 1977. *The Information Economy: Definition and Measurement.* Office of Telecommunications Special Publication 77-12(i), Washington, D.C.: May.

Rappaport, Alfred, ed. 1982. *Information for Decision Making.* Englewood Cliffs, N.J.: Prentice-Hall.

Reich, Robert B. 1983. "The Next American Frontier." *Atlantic Monthly* (March): 43-58, (April): 97-108.

Richardson, Lewis. *Arms and Insecurity.*

Richardson, Peter R., and John R.M. Gordon. 1980. "Measuring Total Manufacturing Performance." *Sloan Management Review* (Winter).

Robinson, A.L. 1980. "Are VLSI Microcircuits Too Hard to Design?" *Science* 209 (July 11).

Rodriguez, Carlos A. 1975. "Trade in Technological Knowledge and the National Advantage." *Journal of Political Economy* 83 (February): 121-135.

Rosenberg, Nathan. 1972. *Technology and American Economic Growth.* White Plains, N.Y.: M.E. Sharpe.

Rothstein, Robert. 1979. *Global Bargaining: UNCTAD and the Quest for a New International Economic Order.* Princeton, N.J.: Princeton University Press.

Sahal, Devendra. 1977. "A Theory of Measurement of Technological Change." *International Journal of Systems Science* 8.

Sahal, Devendra. 1983. "Invention, Innovation and Economic Evolution." *Journal of Technological Forecasting and Social Change.*

Sant, Roger W. "The Least Cost Energy Strategy." The Energy Productivity Center, Mellon Institute, Arlington, Va.

Savitz, Maxine. 1981. "The Current U.S. National Energy Conservation Program in the Residential and Commercial Sectors." In J.D. Millhone and E.H. Willis, eds., *New Energy Conservation Technologies and Their Commercialization*, vol. 1. Berlin: Springer-Verlag.

Shackson, Richard H., and H. James Leach. 1980. "Using Fuel Economy and Synthetic Fuels to Compete with OPEC Oil." Energy Productivity Center, Mellon Institute, Arlington, Va.

Schmitt, Roland W. 1981. "The 1980 Robens Coal Science Lecture: Coal Based Electricity in the United States." *Journal of the Institute of Energy* 54, no. 419.

Schumacher, E.F. 1973. *Small Is Beautiful: Economics as if People Mattered.* New York: Harper & Row.

Schmookler, Jacob. 1966. *Invention and Economic Growth.* Cambridge, Mass.: Harvard University Press.

Schmookler, Jacob J. 1952. "The Changing Efficiency of the American Economy, 1869–1938." *Review of Economics and Statistics* (August).

Schultze, W., and R. G. Cummings. 1979. "Does Conservation of Mass-Energy Matter for Economic Growth?" Working paper, University of New Mexico, June.

Schumpeter, Joseph A. 1939. *Business Cycles.* New York: McGraw-Hill.

Schumpeter, Joseph A. 1950. *Capitalism, Socialism, and Democracy*, 3rd ed., New York: Harper & Row.

Serrin, William. 1973. *The Company and the Union.* New York: Alfred Knopf.

Shane, Harold G. 1982. "The Silicon Age and Education." *Phi Delta Kappan* (January): 303–308.

Sharlin, Harold I. 1967. "Applications of Electricity." In M. Kranzberg and C. Pursell, eds., *Technology in Western Civilization.* New York: Oxford University Press.

Sharlin, Harold I. 1967. "Electrical Generation and Transmission." In M. Kranzberg and C. Pursell, eds., *Technology in Western Civilization.* New York: Oxford University Press.

Simon, Herbert. 1977. *The New Science of Management Decisions.* Englewood Cliffs, N. J.: Prentice-Hall.

Smith, Cyril Stanley. 1967. "Mining and Metallurgical Production, 1800–1880." In M. Kranzberg and C. Pursell, eds., *Technology in Western Civilization.* New York: Oxford University Press.

Smith, R. A. 1966. *Corporations in Crises.* Garden City, N. Y.: Anchor Books, Doubleday, Chapters 3 and 4.

Solow, R. 1957. "Technical Change and the Aggregate Production Function." *Review of Economics and Statistics* (August).

Soma, John T. 1976. *The Computer Industry: An Economic Legal Analysis of Its Technology and Growth.* Lexington, Mass.: Lexington Books.

Stanford Research Institute. 1970. "A Study of Trends in the Demand for Information Transfer." Menlo Park, California: SRI.

Stein, Herbert. 1982. "The Industrial Economies: We Are Not Alone." *The American Enterprise Institute Economic* (May).

Stiglitz, J. 1979. "Neoclassical Analysis of the Economics of Natural Resources." In V. K. Smith, ed., *Scarcity and Growth Reconsidered.* Baltimore: Johns Hopkins University Press.

Stobaugh, Robert, and Daniel Yergin. 1979. *Energy Future: Report of the Energy Project at the Harvard Business School.* New York: Random House.

Strandh, Sigurd. 1979. *A History of the Machine.* New York: A&W Publishers.

"Technology and the Appropriate Theory of the Multinational Corpoation." 1977. In J. Bhagwati, ed., *The New International Economic Order: The North-South Debate.* Cambridge, Mass.: MIT Press.

Terleckyi, Nestor. 1959. "Sources of Productivity Advance." Ph.D. thesis, Columbia University, New York.

Terleckyi, Nestor E. 1974. "Recent Findings Regarding the Contribution of Industrial R&D to Economic Growth." Colloquium on the Relationships between R&D and Economic Growth/Productivity, November 9, 1977, Washington, D.C.: National Science Foundation.

The Protestant Ethic and the Spirit of Capitalism (translated by Talcott Parsons). 1930. London: G. Allen Unwin.

Thurow, Lester. 1981. "Solving the Productivity Problem." In *Alternatives for the 80's, vol. 2 Strengthening the Economy: Studies in Productivity.* Washington, D.C.: Center for National Policy.

Tucker, William. 1980. "The Wreck of the Auto Industry." *Harpers* (November).

Usher, Abbott Payson. 1967. "The Textile Industry, 1750–1830." In M. Kranzberg and C. Pursell, eds., *Technology in Western Civilization.* New York: Oxford University Press.

Utterback, James M., and William J. Abernathy. 1979. "A Dynamic Model of Process and Product Innovation." *Omega* 3.

Veblen, Thorstein. *Imperial Germany and the Industrial Revolution.*

Vernon, Raymond. 1966. "International Investment and International Trade in the Product Cycle." *Quarterly Journal of Economics* (May) 290–207.

Vernon, Raymond. 1970. *Sovereignty at Bay.* New York: Basic Books.

Weber, Max. 1950. *General Economic History.* Glencoe, Ill.: The Free Press.

Weil, Henry. 1982. "Getting Comfortable with Computers." *US Air* (October).

Wells, Louis T. 1969. "Tests of a Product Cycle Model of International Trade: United States Exports of Consumer Durables." *QTE* (February).

Wells, Louis T. 1972. "The Product Life Cycle and International Trade." Harvard Business School, Division of Research, Boston, Massachusetts.

Whitman, Marina V.N. 1981. "International Trade and Investment: Two Perspectives." Frank D. Graham Memorial Lecture, March 5.

Williamson, Harold F. 1967. "Mass Production for Mass Consumption." In M. Kranzberg and C. Pursell, eds., *Technology in Western Civilization.* New York: Oxford University Press.

Winger, John G. et al. 1972. "Outlook for Energy in the United States to 1985." Chase Manhattan Bank, New York.

Winks, Robin W., ed. 1963. *British Imperialism: Gold, God, Glory.* European Problem Studies. Hinsdale, Ill.: Dryden Press.

Woodbury, Robert S. 1960. "The Legend of Eli Whitney and Interchangeable Parts." In *Technology and Culture* (Summer): 235–253.

NAME INDEX

263

ORGANIZATION INDEX

SUBJECT INDEX

ABOUT THE AUTHOR

Robert U. Ayres has been studying technology from many perspectives for over twenty years. Academically trained as a theoretical physicist (Ph.d., 1958), he left academic physics temporarily in 1961 to join the newly formed Hudson Institute where, he undertook the study of potential environmental consequences of nuclear weapons. He also became increasingly fascinated with "future" studies and technological change and began to work on his first book *Technological Forecasting and Long Range Planning* (1969). From the Hudson Institute he moved in 1967 to Resources for the Future, Inc., in Washington, D.C., where he worked on several studies of the environmental consequences of materials/energy transformation and collaborated on two more books on economics, technology, and the environment.

From 1968 through 1978 he continued working and writing intensively in these areas, primarily as a cofounder and vice president of International Research and Technology Corp. (IR&T)—a Washington-based consulting firm.

In January 1979 Dr. Ayres was appointed professor in the Department of Engineering and Public Policy at Carnegie-Mellon University, where his research interests have expanded further to include the economic implications of robotics, computers, and new developments in manufacturing technology. *The Next Industrial Revolution* is his eighth book.